BASIC thermodynamics and heat transfer

BASIC thermodynamics and heat transfer

D H Bacon MSc, BSc(Eng), CEng MIMechE
Department of Mechanical Engineering
Plymouth Polytechnic

Butterworths
London . Boston . Durban . Singapore . Sydney . Toronto . Wellington

First published, 1983

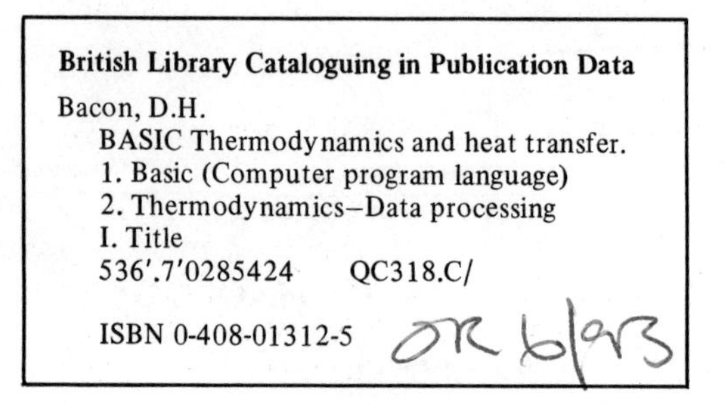
British Library Cataloguing in Publication Data

Bacon, D.H.
BASIC Thermodynamics and heat transfer.
1. Basic (Computer program language)
2. Thermodynamics–Data processing
I. Title
536'.7'0285424 QC318.C/

ISBN 0-408-01312-5

Typeset by Scribe Design, Gillingham, Kent
Printed and bound by Whitstable Litho Ltd., Whitstable, Kent

Preface

Computing is an important tool for an engineer and the linking of computing with an engineering subject in a single book will help the engineer to develop skill in using this tool. BASIC is a widespread, simple language to learn and use, which rapidly engenders confidence and is available for most microcomputers.

BASIC thermodynamics and heat transfer is not intended as a comprehensive treatise on BASIC or thermodynamics and heat transfer. Instead it aims to help readers to become proficient at BASIC programming by using it in an engineering subject and to use computing as an aid to understanding the subject. In particular the advantage of being able to study the effect on the results of the program of changing the values of the input data must convince the engineer of the value of computing for design and optimisation calculations.

Chapter 1 introduces BASIC. Chapter 2 outlines the fundamental thermodynamics normally associated with undergraduate courses as does the work on properties in Chapter 3, process and cycle analysis in Chapter 4 and fluid flow topics in Chapter 5.

When these preliminary sections have been completed it will be found that Chapter 6 consists of a series of examples and problems on particular applications in which any necessary extra theory is included in the question text. Chapter 7 is entirely devoted to heat transfer and could be used separately from Chapters 2 to 6. Indeed, some readers may concentrate on this section alone.

Others may prefer to examine all the examples and problems relevant to a particular application, for which the table below will assist.

	Examples	*Problems*
Steam plant	3.3, 3.5, 4.1, 6.1, 6.2	3.1, 4.1, 4.4, 5.2, 6.1, 6.2
Refrigeration	3.4, 4.4, 6.5	3.2, 3.3, 4.8, 6.8
Gas turbines	4.3, 5.1, 5.2, 6.3	4.3, 5.3, 6.3, 6.4, 6.5
Air conditioning	3.7, 6.8	3.5, 4.7, 6.11
Reciprocating compressors	6.7	4.5, 6.9
Reciprocating IC engines	6.4, 6.9	4.6, 6.6, 6.7, 6.10

In whatever manner this book is used it will be found that the examples are reasonably simple and most answers can be checked easily. The programming is also simple but may appear to the experienced reader to lack elegance. However, elegant programs are often difficult for the beginner to understand and it is intended that the simple approach will avoid difficulties which might defeat the object of the book.

Acknowledgement is due to Don Goodsell for persistence and to Brian Selley, Michael Iremonger and Peter Smith for assistance.

D.H.B

Contents

Chapter 1

Introduction to BASIC

1.1 The BASIC approach

The programs in this book are written in the BASIC programming language. This book is not an instruction manual in BASIC but a short description of the simple BASIC used in the book follows.

1.2 The elements of BASIC

1.2.1 Mathematical expressions

One of the main objects of the example programs in this book is to assist in the learning of BASIC by applying it to a relevant engineering subject. This aim can be met by the reader studying the examples, possibly copying them and then trying some of the problems. It will be necessary to evaluate equations which contain numerical constants, variables (e.g. x) and functions (e.g. sine). All numbers are treated identically whether they are integer (e.g. 36) or real (e.g. 36.1). An exponential form is used to represent large or small numbers (e.g. 3.61E6, which equals 3.61×10^6). Numeric variables are represented by a letter, or a letter followed by a digit (e.g. T or T1). Some computers allow two letters, or three letters plus two figures, etc, for variables, which makes representation easier for the user. For generality π is always written as 3.142 in this book. The determination of square roots uses an in-built function (e.g. SQR(X)). The argument in brackets (X) can be a number, a variable or a mathematical expression. For trigonometric functions (SIN(X), COS (X), etc.) the argument is interpreted as being measured in radians. Other functions include a natural logarithm (LOG) and its exponential (EXP) and ABS which gives the absolute value of the argument.

Mathematical equations also contain operators such as plus and minus, etc. These operators have a hierarchy in that some are performed by the computer before others. In descending order of priority the operators are

to the power of (↑)
multiply (*) and divide (/)
add (+) and subtract (–)

Thus, for example, multiplication is done before addition. The computer works from left to right if the operators have the same hierarchy. Brackets can be used to override any of these operations. Hence

$$\frac{a+b}{3c} \text{ becomes } (A + B)/(3*C) \text{ or } (A + B)/3/C$$

1.2.2 Program structure and assignment

A BASIC program is a sequence of statements which define a procedure for the computer to follow. As it follows this procedure the computer allocates values to each of the variables. The values of some of these variables may be specified by data that is input to the program. Others are generated in the program using, for instance, the assignment statement. This has the form

line number [LET] variable = mathematical expression

where the word LET is usually optional and therefore omitted. For example, the root of a quadratic equation

$$x_1 = \frac{-b + \sqrt{b^2 - 4ac}}{2a}$$

may be obtained from a statement such as

```
100 X1 = (–B + SQR (B ↑ 2 – 4*A*C))/(2*A)
```

It is important to realise that an assignment statement is not itself an equation. It is an instruction to give the variable on the left-hand side the numeric value of the expression on the right-hand side. Thus it is possible to have a statement

```
50 X = X + 1
```

which increases by 1 the value of X. Each variable can have only one value at any time unless it is subscripted (see section 1.2.7).

Note that all BASIC statements (i.e. all the program lines) are numbered. This defines the order in which they are executed.

1.2.3 Input

For interactive or conversational programs the user specifies values for variables in response to prompts from the computer while the program is running. The statement has the form

line number INPUT variable 1 [, variable 2, . . .]

for example

```
20 INPUT A, B, C
```

When the program is run the computer prints ? as it reaches this statement and waits for the user to type values for the variables:

```
? 5 ? 10 ? 15
```

which makes A = 5, B = 10 and C = 15 in the above example.

An alternative form of data input is useful if there are many data or if the data are not to be changed by the user (e.g. the steam table data). For this type of data specification there is a statement of the form

line number READ variable 1 [, variable 2, . . .]

for example

```
20 READ A, B, C
```

with an associated statement (or number of statements) of the form

line number DATA number 1 [, number 2, . . .]

for example

```
1 DATA 5, 10, 15
```

or

```
1 DATA 5
2 DATA 10
3 DATA 15
```

Data statements can be placed anywhere in the program; it is often convenient to place them at the beginning or end so that they can be easily changed.

1.2.4 Output

Output of data and the results of calculations, etc. is done using a statement of the form

line number PRINT list

The list may contain variables or expressions such as

```
200 PRINT A, B, C, A*B/C
```

text enclosed in quotes,

```
10 PRINT "INPUT A, B, C IN MM";
```

or mixed text and variables

```
300 PRINT "PRESSURE = "; P; "KN/M↑2"
```

The items in the list are separated by commas or semicolons. Commas give tabulation in columns, each about 15 spaces wide. A semicolon suppresses this spacing, and if it is placed at the end of a list it also suppresses the line feed. If the list is left unfilled a blank line is printed.

Note the necessity to use PRINT statements in association with both 'run-time' input (to indicate what input is required) and READ/DATA statements (because otherwise the program user has no record of the data).

1.2.5 Conditional statements

It is often necessary to enable a program to take some action if, and only if, some condition is fulfilled. This is done with a statement of the form

line number IF expression 1 conditional operator expression 2 THEN GOTO line number

where the possible conditional operators are

=	equals
<>	not equal to
<	less than
<=	less than or equal to
>	greater than
>=	greater than or equal to

For example, a program could contain the following statements if it is to stop when a zero value of A is input

```
20 INPUT A
30 IF A<>0 THEN GOTO 50
40 STOP
50 . . .
```

Note the statement

line number STOP

which stops the run of a program.

1.2.6 Loops

There are several means by which a program can repeat some of its actions or calculations. The simplest such statement is

line number GOTO line number

This could be used, for instance, with the conditional statement example above so that the program continues to request values of A until the user inputs zero.

The most common means for performing loops is with a starting statement of the form

line number FOR variable = expression 1 TO expression 2 [STEP expression 3]

where the STEP is assumed to be unity if omitted. The finish of the loop is signified by a statement

line number NEXT variable

where the same variable is used in both FOR and NEXT statements. Its value should not be changed in the intervening lines.

A loop is used if, for example, N sets of data have to be READ and their reciprocals printed:

```
10 READ N
20 PRINT "NUMBER", "RECIPROCAL"
30 FOR I = 1 TO N
40 READ A
50 PRINT A, 1/A
60 NEXT I
```

1.2.7 Subscripted variables

It is sometimes useful to allow a single variable to have a number of different values during a single program run. For instance, if a program contains data for several materials it is convenient for their densities to be called R(1), R(2), R(3), etc. instead of R1, R2, R3, etc. It is then possible for a single statement to perform calculations for all the materials:

```
50 FOR I = 1 to N
60 M(I) = V * R(I)
70 NEXT I
```

which determines the mass M(I) for each material from the volume (V) of the body. Some computers print subscripted variables with square brackets which are easily distinguished.

A non-subscripted variable has a single value associated with it but if a subscripted variable is used it is necessary to provide space in the computer's memory for all the values. This is done with a dimensioning statement of the form

line number DIM variable 1 (integer 1) [, variable 2 (integer 2), . . .]

For example

20 DIM R(50), M(50)

which allows up to 50 values of R and M. The DIM statement must occur before the subscripted variables are first used so it is usual to find it very early in the program.

1.2.8 Subroutines

Sometimes a sequence of statements needs to be accessed more than once in the same program. Instead of merely repeating these statements it is better to put them in a subroutine. The program then contains statements of the form

line number GOSUB line number

When the program reaches this statement it branches (i.e. transfers control) to the second line number. The sequence of statements comprising the subroutine and starting with this second line number ends with a statement

line number RETURN

and the program returns control to the statement immediately after the GOSUB call.

Subroutines can be placed anywhere in the program but it is usually convenient to position them at the end, separate from the main program statements.

1.2.9 Other statements

Explanatory remarks or headings which are not to be output can be inserted into a program using

line number REM comment

Any statement beginning with the word REM is ignored by the computer.

The integer part of a number can be derived by the INP function. The example below will also round up the printout.

line number variable = (INP (variable * 1000 + 0.5))/1000

The multiplier 1000 (and the divisor) may need to be 10 to any power depending on the magnitude of the variable.

1.3 Checking programs

Most computers give a clear indication if there are grammatical (syntax) errors in a BASIC program. Program statements can be modified by retyping them correctly or by using special editing procedures. The majority of syntax errors are easy to locate but if a variable has been used with two (or more) different meanings in separate parts of the program some mystifying errors can result.

It is not sufficient for the program to be just grammatically correct; it must also give the correct answers. A program should therefore be checked either by using data which gives a known solution or by hand calculation. If the program is to be used with a wide range of data, or by users other than the program writer, it is necessary to check that all parts of it function. It is also important to ensure that the program does not give incorrect but plausible answers when 'nonsense' data is input. It is quite difficult to make programs completely 'user-proof' and they often become lengthy in being made so. The programs in this book have been kept as short as possible for the purpose of clarity and may not therefore be fully 'userproof'.

1.4 Summary of BASIC statements

Statement	Description
Assignment	
DIM	Allocates space for subscripted variables
Input	
INPUT	'Run-time' input of data from keyboard
READ	Reads data from DATA statements
DATA	Storage area for data
Output	
PRINT	Prints output
Program control	
STOP or END	Stops program run
GOTO	Unconditional branching
IF . . THEN GOTO	Conditional branching
FOR . . TO . . STEP . .	Opens loop

NEXT	Closes loop
GOSUB	Transfers control to subroutine
RETURN	Returns control from subroutine
Comment	
REM	Comment in program
Functions	
SQR	Square root
SIN	Sine (angle in radians)
COS	Cosine (angle in radians)
LOG	Natural logarithm (base e)
EXP	Exponential
ABS	Absolute value
INP	Integer part (in some versions of BASIC this is the INT function)

1.5 References

1. Kemeny, J.G. and Kurtz, T.E., *BASIC Programming,* John Wiley (1968).
2. Monro, D.M., *Interactive Computing with BASIC,* Edward Arnold (1974).
3. Alcock, D., *Illustrating BASIC,* Cambridge University Press (1977).

Chapter 2

Thermodynamics

ESSENTIAL THEORY

2.1 Fundamental concepts

Engineering thermodynamics is mainly concerned with energy transfers within the working substance used in a machine. The machine may be designed to produce power by burning fuel or to consume power whilst raising pressure or to cause a heat transfer, etc. Whatever the purpose the method of approach to a problem is to separate the *system* to be investigated from the *surroundings* with a *boundary* and to apply the laws of thermodynamics. To solve the resulting equations other information may be required concerning, for example, fluid flow or heat transfer, together with a knowledge of the *property* changes involved in the *processes* used in the machine.

A property of a system is any observable characteristic (pressure, volume, temperature, energy, etc.)*. The *equilibrium state* of the system is defined by properties. When the state changes by means of some process the change of state is defined by the change in the properties. Two states will be identical if the value of every property is identical. For simple working substances known as *pure substances* the *two-property rule* states that any two independent properties are adequate to define a state. Air and steam are examples of pure substances. Since only two properties are needed to define state, graphs and tables can be used to display property values.

Processes may be *reversible* or *irreversible.* Reversible processes may be represented by a line on a two-property equilibrium graph. This line implies a continuous series of equilibrium states, and represents an idealised process, which may be analysed mathematically. The term reversible stems from the fact that to move along such a line continuously in equilibrium would involve changes of state brought about by infinitesimally small gradients. If these gradients were *reversed* the process should reverse its path and return to its initial state. This means that there must be no losses due to friction or any other cause during the

*For a full list of properties, see section 3.2.

process. It is obvious that the reversible process is an ideal never achieved in practice. It is analogous to the concepts in mechanics of frictionless pulleys and particles without mass.

Practical processes are not a series of equilibrium states and often little is known about the process which occurs between two known end states. Such processes are termed *irreversible processes.* They cannot be meaningfully represented on two-property equilibrium graphs and do not lend themselves to mathematical analysis. Practical processes are therefore investigated by forming a model with the equivalent idealised reversible process and then mutliplying or dividing by a factor called a *process efficiency.*

A *cycle* is a series of processes whose beginning and end states are identical. The process paths of a reversible cycle on a two-property graph will form a closed figure.

A change from one equilibrium state to another may be accompanied by a boundary interaction. This interaction only appears during the change of state and is associated with an energy transfer. Two kinds of interaction occur:

(a) energy transfer by work (work transfer);
(b) energy transfer by heat (heat transfer).

Work and heat are not present in a system and are not properties. They only appear as methods of energy transfer during a change of state.

2.2 Work transfer and power

In simple terms there are two forms of work transfer. They are called displacement work transfer and shaft work transfer.

If a system boundary deforms under the action of the system pressure acting on the boundary, the state of the system will change and a *displacement work transfer* occurs. Consider a small element of boundary of area δA which moves δL under the action of the pressure p. Displacement work transfer

$$\delta W = p\delta A \ \delta L = p\delta V_{\mathrm{sw}}$$

where δV_{sw} is the volume swept by the small element of boundary.

Summing for the whole system boundary

$$W = \Sigma\, \delta W = \Sigma p\, \delta V_{\mathrm{sw}}$$

For the reversible case in which there is a continuous relationship between the property p and the property V

$$W = \int p \, \mathrm{d}V \tag{2.1}$$

This mathematical relationship enables the work transfer to be evaluated for any reversible process in which there is a known continuous relation between p and V. It can also be seen that if an equilibrium pressure–volume diagram were drawn and the reversible process represented upon it, then the area under the process path line represents the work transfer.

Work transfer is also achieved by means of a rotating shaft. In this case the term *shaft work transfer* is used. For example, there is shaft work transfer from a turbine to a compressor in a gas turbine engine. (The term *stirring work transfer* is sometimes used in the less common case where a fan or stirrer is used as a method of energy tranfer to a system.) In reversible shaft work

$$W_x = \int T \, d\theta$$

where T is the torque and θ is the angular distance moved by the torque.

In most practical applications work transfer occurs continuously and the engineer is interested in the *work transfer rate* or *power* produced or absorbed by a system. The cycle is repeated at regular intervals and the power is obtained by multiplying the work transfer per cycle by the cyclic frequency.

2.3 Heat transfer

Heat transfer occurs due to a difference in temperature. If two systems at differing temperatures are brought into contact the interaction occurs until thermal equilibrium (equality in temperature) is reached. The symbol for heat transfer is Q. In idealised situations heat transfers are sometimes associated with reservoirs. A reservoir remains at a fixed uniform temperature irrespective of any heat transfer.

2.4 The first law of thermodynamics

This is a law of energy conservation and when applied to a system which undergoes a process is written

$$Q - W = \Delta E \quad \text{or} \quad q - w = \Delta e$$

where Q = heat transfer (q = specific heat transfer, Q/m)
W = work transfer (w = specific work transfer, W/m)
ΔE = energy change (Δe = specific energy change, $\Delta E/m$)
m = mass of working substance in the system
and Δ = 'change in', in the sense (final value – initial value).

In this form the equation has an inbuilt sign convention that *heat transfer* to *a system is positive and work transfer* from *a system is positive.*

When applied to a cycle, ΔE is zero (as energy is a property) and the first law becomes

$$\Sigma Q = \Sigma W$$

If a system is insulated so that no heat transfer can occur the system and process are called *adiabatic*, and

$$-W = \Delta E$$

In mechanical engineering, three forms of energy are usually considered to constitute E. These are internal energy due to state (U), kinetic energy ($\frac{1}{2}mV^2$) and potential energy (mgz).

$$E = U + \tfrac{1}{2} mV^2 + mgz$$

or

$$e = u + \tfrac{1}{2}V^2 + gz$$

where $u = U/m$.
Thus the first law becomes (in a specific form)

$$q - w = \Delta(u + \frac{V^2}{2} + gz)$$

2.5 The non-flow energy equation

Many systems are at rest relative to the observer and also have very small changes in potential energy. Such systems are called non-flow and involve non-flow processes. The first law becomes the *non-flow energy equation*

$$Q - W = \Delta U \text{ or } q - w = \Delta u \qquad (2.2)$$

The most common applications of this equation are to systems contained by piston and cylinder mechanisms, and to problems involving filling rigid vessels with a substance from a constant supply state (Figure 2.1).

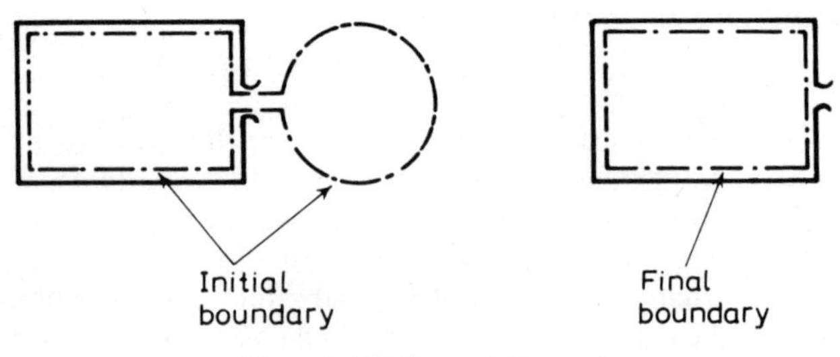

Figure 2.1 Filling a rigid vessel

In the latter case

$$Q - W = U_{\text{final}} - U_{\text{initial}} \tag{2.3}$$

U_{initial} in the case shown in Figure 2.1 will have two terms, one for the vessel contents and one for the bubble about to enter the vessel as work is done on the bubble.

2.6 The flow energy equations

In most practical machines the working substance flows continuously through the device and shaft power is continuously produced or absorbed. The processes are called flow processes and the concept of the system is modified to allow for the motion relative to the observer.

The method is to consider flow through a fixed surface in space called a *control surface.* The control surface is chosen to enclose fluid that is passing through the machine (or part of the machine) under consideration. The volume enclosed is called the *control volume.* It is of fixed shape, size and position relative to the observer.

Consider a control surface enclosing a control volume. Let there be a working substance entering the control volume at station 1 and leaving the control volume at station 2. Kinetic energy and potential energy may not be negligible in this situation.

The first law of thermodynamics is applied to a system consisting of the mass within the control volume m_{cv} and a small amount of mass just about to enter. During a small time interval the system boundary will deform as the mass enters the control volume and a similar small mass leaves. During this time interval the shaft work transfer W_{x} and the heat transfer Q are effected. The displacement or boundary deformation work is not included in the W_{x} term. The total work transfer is therefore

$$W = W_{\text{x}} + \text{displacement work transfer}$$

The resulting equation is

$$\dot{Q} - \dot{W}_{\text{x}} = \left(\frac{\mathrm{d}E}{\mathrm{d}\tau}\right)_{\text{cv}} + \left(\Sigma_{\text{out}} - \Sigma_{\text{in}}\right) \dot{m} \left(u + pv + \frac{V^2}{2} + gz\right)$$

where $\left(\frac{\mathrm{d}E}{\mathrm{d}\tau}\right)_{\text{cv}}$ is the rate of change of energy stored in the control volume, $(\Sigma_{\text{out}} - \Sigma_{\text{in}})$ means the sum of all outgoing stream contributions less the sum of all ingoing stream contributions, $\dot{Q}$ is the heat transfer rate, $\dot{W}_{\text{x}}$ is the shaft work transfer rate or power, and $\dot{m}$ is the mass flow rate of working substance.

It is usual to write $(u + pv)$ as h, the specific enthalpy. Enthalpy is itself a property, being made up of other properties.

For most applications $(\mathrm{d}E/\mathrm{d}\tau)_{\mathrm{cv}}$ is zero, which means that the flow is steady, and in many applications there is only a single stream entering and leaving at a steady flow rate $\dot{m}$, so that the *steady flow energy equation* applied to steady flow processes becomes

$$\dot{Q} - \dot{W}_{\mathrm{x}} = \dot{m}\ \Delta(h + \frac{V^2}{2} + gz)$$

or

$$q - w_{\mathrm{x}} = \Delta(h + \frac{V^2}{2} + gz) \qquad (2.4)$$

The value of the mass flow rate m is sometimes obtained from the continuity equation

$$\dot{m} = \rho A V = \text{constant} \qquad (2.5)$$

where ρ = fluid density at the point where the flow area *normal* to the flow velocity V is A.

It will be found that in many applications the change in specific enthalpy Δh is large compared with the change in kinetic energy and potential energy and the steady flow energy equation then becomes

$$q - w_{\mathrm{x}} = \Delta h \quad \text{or} \quad \dot{Q} - \dot{W}_{\mathrm{x}} = \dot{m}\ \Delta h$$

2.7 The second law of thermodynamics

The first law of thermodynamics states that energy is conserved; the second law places constraints on the way in which the conservation is permitted to apply.

The machine shown in Figure 2.2(a) does not contravene the first law since energy is conserved. However all attempts to build such a

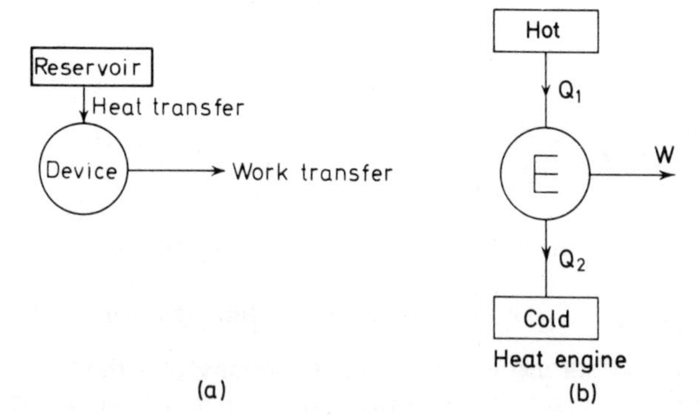

Figure 2.2

machine have failed. As a result the second law of thermodynamics has been formulated: *It is impossible to construct a device which will operate in a cycle and produce, as its sole effect, a positive work transfer whilst using a single reservoir for heat transfer.* It is found, by using a second reservoir to which there is heat transfer from the cycle, that a continuously operating system producing work transfer can be constructed. Figure 2.2(b) shows such a machine. It is called a *heat engine.* A steam plant is a practical example of a heat engine. The first law is satisfied since

$$Q_1 - Q_2 = W$$

but unfortunately the work output is less than the heat input.

The *thermal efficiency* of the machine quantifies the amount of work obtained

$$\eta_{th} = \frac{W}{Q_1} \tag{2.6}$$

Further work with the second law enables this efficiency to be evaluated for an ideal reversible heat engine in terms of the thermodynamic (absolute) temperatures of the reservoirs

$$\eta_{th} = \frac{T_{max} - T_{min}}{T_{max}}$$

The sole significance of this expression for the Carnot efficiency is that it determines the maximum possible efficiency that can be achieved by a reversible heat engine operating between two temperature levels. It therefore offers an (unachievable) target for the engineer.

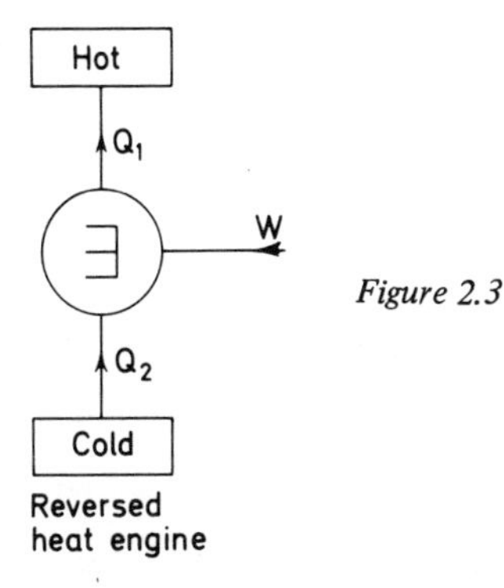

Figure 2.3

Similar concepts can be applied to a *reversed heat engine* (Figure 2.3) to which an alternative statement of the second law may be applied: *It is impossible to construct a device which will operate in a cycle and effect heat transfer from a cold reservoir to a hot reservoir as its sole effect.* A reversed heat engine transfers heat from a cold reservoir to a hot reservoir but needs a work input to achieve this. Such a machine is called a *refrigerator* or a *heat pump,* and for these devices the equivalent of thermal efficiency is the *coefficient of performance* (c.o.p.). For a refrigerator

$$\text{c.o.p.} = \frac{Q_2}{W}$$

which ideally becomes

$$\frac{T_{min}}{T_{max} - T_{min}} \tag{2.7}$$

For a heat pump

$$\text{c.o.p.} = \frac{Q_1}{W}$$

which ideally becomes

$$\frac{T_{max}}{T_{max} - T_{min}} \tag{2.8}$$

Another concept which derives from the second law is that of a property, *entropy* S, which may be defined: *Entropy is the property of a system whose change in a given process is equal to* $\Sigma(\delta Q/T)$ *evaluated in a reversible process between the same two end states as those of the given process,* where δQ is a small heat transfer taking place at temperature T. This may be rearranged to yield

$$Q = \int T\mathrm{d}S \quad \text{or} \quad q = \int T\mathrm{d}s \text{ (where } s = S/m) \tag{2.9}$$

which leads to the useful concept of a $T - S$ property graph on which the area under a *reversible* process path will represent heat transfer in a similar way to the $p - V$ graph and work transfer.

The application of entropy is not easy but for an adiabatic process (which is important because many real machines use processes which are approximately adiabatic) it can be shown that

$$\Delta S = 0 \text{ if the process is reversible}$$

$$\text{and } \Delta S > 0 \text{ if the process is irreversible}$$

This can be reorganised to give (see section 2.8) an adiabatic flow process efficiency (isentropic efficiency). It is also possible by means of the second law to show that all real heat transfer processes are irreversible since they cause entropy increases in adiabatic systems. There are many other applications of this fundamental law of nature which are not relevant at this stage.

2.8 Isentropic flow process efficiency

In this efficiency energy transfer in an ideal reversible adiabatic process is compared with that in a real irreversible adiabatic process. Three cases are of interest.

(1) Work producing process (Figure 2.4(a)).

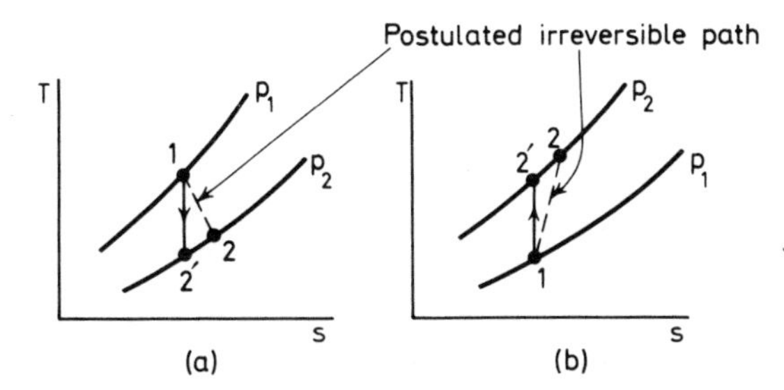

Figure 2.4 Isentropic efficiency

In a turbine the actual expansion process 1–2 shows an increase in entropy but the ideal process 1–2′ shows constant entropy. The ideal work can be evaluated and the actual work can then be obtained from:

$$\eta_{isen} = \frac{\text{actual work}}{\text{ideal work}}$$

By use of the steady flow energy equation this becomes (provided changes in kinetic and potential energy are negligible)

$$\eta_{isen} = \frac{\text{actual work}}{\text{ideal work}} = \frac{h_1 - h_2}{h_1 - h_2{}'} \tag{2.10}$$

(2) Work absorbing process (Figure 2.4(b)).
In a compressor the actual compression process 1–2 shows an increase in entropy whereas the ideal process 1–2′ has constant entropy. By similar arguments to those for the expansion process it may be shown that

$$\eta_{isen} = \frac{\text{ideal work}}{\text{actual work}} = \frac{h_2{}' - h_1}{h_2 - h_1} \tag{2.11}$$

(3) Kinetic energy producing process (Figure 2.4(a)).

In a nozzle the actual expansion process 1–2 shows an increase in entropy whereas the ideal process 1–2′ has constant entropy. Thus by similar arguments it may be shown that:

$$\eta_{\text{isen}} = \frac{\text{actual K.E. produced}}{\text{ideal K.E. produced}} = \frac{h_1 - h_2}{h_1 - h_2{}'} \tag{2.12}$$

2.9 *T*–d*s* relations

A useful relationship between the property entropy produced from the second law and the more easily visualised properties pressure, temperature and energy can be derived. This shows that

$$T\text{d}s = \text{d}u + p\text{d}v$$

or

$$T\text{d}s = \text{d}h - v\text{d}p \tag{2.13}$$

This may be used to determine entropy changes.

2.10 Summary

This section has set out the concept of the system and energy transfers and has applied the laws of thermodynamics to such a model. In order to be able to solve problems it is now necessary (as suggested in the opening paragraph) to supply information about property values for working substances and details of the ideal processes used to model those commonly used in practical applications.

Chapter 3

Property data

ESSENTIAL THEORY

3.1 Summary

In order to apply the laws of thermodynamics to systems and obtain numerical solutions to problems the physical properties of working substances are required. These are measured by experiment and may be presented as tables, graphs or equations. To use a computer requires that skeleton tables be programmed with a suitable interpolation routine or that equations be programmed to replace a chart. Complex sets of equations are available for common substances such as steam[1] but for less common substances the skeleton table method is needed. This method is used here and data is presented for a limited working range for water/steam and refrigerant 12. Other substances can be treated by forming similar data blocks from tables.

For simple gas calculations it is usual to assume that the ideal gas equation is valid and the resulting equations are then so trivial that they make tabulation unnecessary. Psychrometric mixtures of air and steam are normally associated with charts or tables, and a program is included to solve problems involving these mixtures.

3.2 Properties

A number of properties such as pressure, temperature, volume and energy are available to describe the state of a pure substance. Some of these properties depend on the mass of substance considered, e.g. volume and energy, and are called *extensive* properties whereas others, of which pressure and temperature are examples, are independent of mass and are termed *intensive.* Extensive properties may be divided by the mass with which they are associated and then are expressed specifically. Thus specific volume is volume/mass ($v = V/m$).

The following table shows the properties of interest in thermodynamics and is used to define enthalpy and specific heat capacity as functions of other properties.

Name	*Symbol*	*Unit*	*Notes*
Pressure	p	Pa, N/m², bar	1 bar = 10^5 N/m²
Thermodynamic or absolute temperature	T	K	
Volume	V	m³	
Specific volume	v	m³/kg	$v = 1/\rho$
Density	ρ	kg/m³	$\rho = 1/v$
Internal energy	U	kJ	
Specific internal energy	u	kJ/kg	
Enthalpy	H	kJ	$H = U + pV$
Specific enthalpy	h	kJ/kg	$h = u + pv$
Entropy	S	kJ/K	
Specific entropy	s	kJ/kg K	
Dryness fraction	x		
Specific heat capacity at constant pressure	C_p	kJ/kg K	$C_p = \left(\frac{\partial h}{\partial T}\right)_p$
Specific heat capacity at constant volume	C_v	kJ/kg K	$C_v = \left(\frac{\partial u}{\partial T}\right)_v$
Ratio of specific heat capacities	γ		$\gamma = \frac{C_p}{C_v}$
Specific gas constant	R	kJ/kg K	$R = C_p - C_v$
Molar or universal gas constant	R_0	kJ/kmol K	$R_0 = 8.3143$
Relative molecular mass	M		$MR = R_0$

3.3 Phase change

A working substance in common use is steam. The sequence of events which occurs when unit mass of water is heated at constant pressure may be used to describe the construction of property tables.

As the *compressed liquid* is heated, the temperature T rises and the specific volume v increases until, at the *saturation temperature,* the liquid starts to evaporate and change from the *liquid phase* to the *vapour phase*. During the change of phase T remains constant but v continues to increase rapidly. At the commencement of evaporation the liquid is termed *saturated* and, at the end of evaporation, the vapour is termed *dry saturated.* Between these two clearly defined states there is a mixture of liquid and vapour called a *wet vapour.* After the evaporation process has been completed, further heating results in a rise in T and an increase in v as the vapour becomes *superheated.* The difference between the saturation temperature and the temperature of the superheated vapour is called the *degree of superheat.*

By repeating the experiment at a series of different pressures a complete picture of the behaviour of water/steam is obtained, which shows that, as the pressure is increased, the transition stage between liquid and vapour becomes progressively shorter until the critical temperature is reached. States above the critical temperature are *gaseous.*

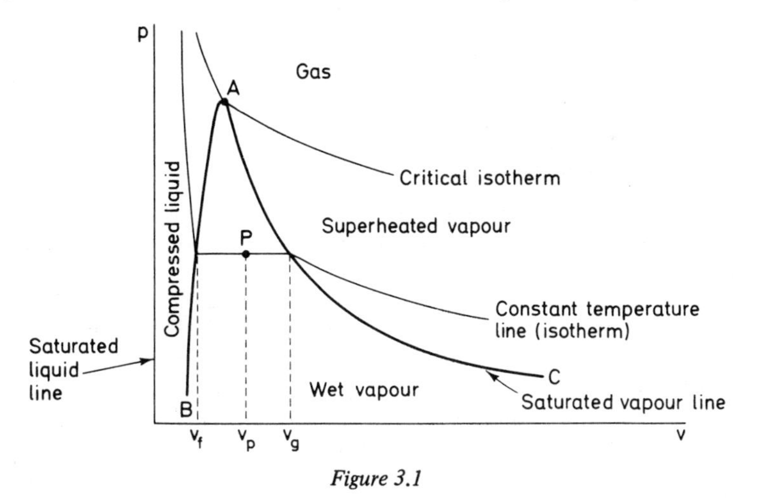

Figure 3.1

If pressure is plotted against specific volume (Figure 3.1) the results of the processes described above are illustrated. The point A is called the *critical point* and no liquid phase can exist at temperatures greater than that of the isotherm through A. This isotherm divides the area for which tabulation is used (properties of vapours) from the area for which the ideal gas rule is used (properties of gases). The locus AB of saturated liquid states is called the *saturated liquid line* and points on it are given the suffix f; the locus AC of dry saturated vapour states is called the *saturated vapour line* and points on it are given the suffix g.

Between states f and g, where pressure and temperature are both constant and not independent, the water is partially vapourized and the properties are defined by the dryness fraction, x given by

$$x = \frac{\text{mass of dry saturated vapour}}{\text{total mass of liquid and vapour}}$$

The mixture at P (Figure 3.1) is made up of x parts of dry saturated vapour and $(1 - x)$ parts of saturated liquid so that the specific volume at P

$$v_p = (1 - x)v_f + xv_g \qquad (3.1)$$

or

$$v_p = v_f + x v_{fg} \quad \text{where } v_{fg} = v_g - v_f$$

Similarly

$$h_p = (1 - x)h_f + x\, h_g$$
$$u_p = (1 - x)u_f + x\, u_g$$
$$s_p = (1 - x)s_f + x\, s_g$$

3.4 Property tabulations and charts

Tables are normally available which give data for saturated liquid and saturated vapour, listing, for a given saturation pressure or temperature, some or all of the quantities v_f, v_g, u_f, u_g, h_f, h_{fg}, h_g, s_f and s_g. The tables enable v, u, h or s to be obtained for saturated liquid, wet vapour and dry saturated vapour and they may also be used for compressed liquid by assuming that u, h and s are functions of temperature only.

Data for superheated vapour is given in a separate table listing some or all of u, h, v (or ρ) and s for an entry of pressure *and* temperature.

One steam chart is readily available which shows h as ordinate against s as abscissa. It is called a Mollier chart and has contours enabling p, T or x to be used to fix a point on the chart from which h may be read off in either superheated, dry saturated or wet states. Similar charts and tables are available for other substances.

3.5 Mathematical equations to property data

Equations giving close approximations to experimental data are available for some substances and may be used for computer generation of property values[1]. Due to the phase behaviour of substances, these equations are different for the various zones and also complex. As an approximation of practical value, the gas zone is treated by simple equations.

3.6 Ideal and perfect gas relations

An ideal gas is described by the equation

$$pv = RT \tag{3.2}$$

where R is the specific gas constant. Since R is constant this may also be written

$$\frac{pv}{T} = \text{constant}$$

and since $v = V/m$, $pV = mRT$.

An ideal gas has variable specific heat capacity, which is not always convenient, and a perfect gas is described as an ideal gas with constant specific heat capacity. This avoids the necessity of having a relation between specific heat and temperature for the evaluation of specific enthalpy, energy and entropy changes in simple calculations. For an ideal gas the definition of specific heat capacity gives

$$\mathrm{d}h = C_p \,\mathrm{d}T$$

which for a perfect gas with constant C_p becomes

$$h_2 - h_1 = C_p(T_2 - T_1) \tag{3.3}$$

Similarly

$$u_2 - u_1 = C_v(T_2 - T_1) \tag{3.4}$$

and

$$s_2 - s_1 = C_p \ln\left(\frac{T_2}{T_1}\right) - R \ln\left(\frac{p_2}{p_1}\right) \tag{3.5}$$

where suffixes 2 and 1 represent the values of properties at the initial and final conditions.

3.7 Properties of non-reactive mixtures

Two mixture types frequently occur: (a) mixtures of gases; and (b) mixtures of gases with vapours.

(a) Properties of gaseous mixtures may be obtained by the equations listed below. It can be seen that they are based on the additive Gibbs–Dalton law:

> *The pressure, internal energy, enthalpy and entropy of a gaseous mixture are respectively equal to the sums of the pressures, internal energies, enthalpies and entropies, which each component would possess if it alone occupied the volume of the mixture at the temperature of the mixture.*

Let the volume of the mixture be V and the temperature T. The following relations apply to the mixture and its components. The suffix i represents the typical ith component.

$$T = T_i,\ V = V_i,\ mv = m_i v_i$$

$$m = \Sigma m_i,\ \rho = \Sigma \rho_i,\ p = \Sigma p_i$$

$$U = \Sigma U_i = \Sigma m_i u_i$$

$$H = \Sigma H_i = \Sigma m_i h_i$$

$$S = \Sigma S_i = \Sigma m_i s_i$$

It may also be shown that a mixture of ideal gases does itself obey the ideal gas equation with

$$R = \frac{\Sigma m_i R_i}{m}$$

where

$$R = \frac{R_0}{M} \text{ and } M = \frac{m}{\Sigma n_i} \quad (n \text{ is the number of moles})$$

Similarly

$$C_p = \frac{\Sigma m_i C_{p_i}}{m}, \quad C_v = \frac{\Sigma m_i C_{v_i}}{m}$$

(b) Properties of mixtures of gases with vapours may be approached through additive methods using the data from each source. One

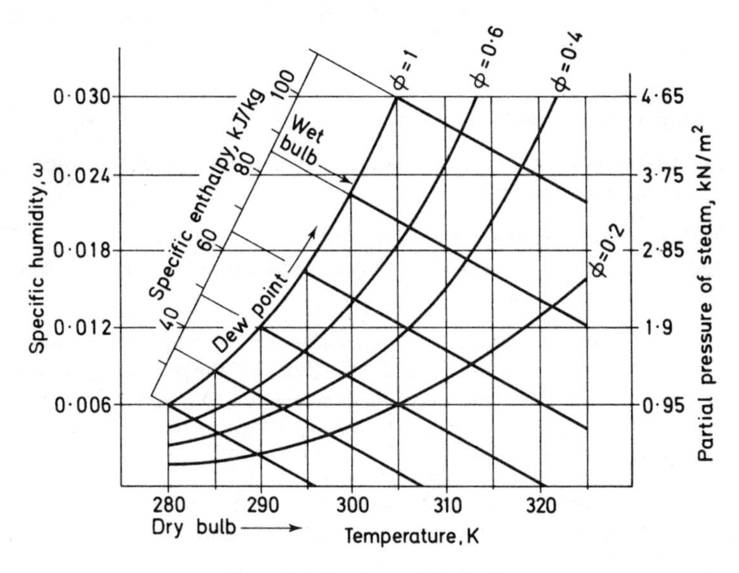

Figure 3.2 Psychrometric chart

particular case of interest is that of atmospheric air. For this special case, tables and charts are available (Figure 3.2) and a program is given for use in problems. The program is based on the work of section 3.8, which also defines the special terminology associated with psychrometry.

3.8 Psychrometry

For the derivation of atmospheric air properties the following terms and concepts are used. Atmospheric air consists of a mixture of dry air and superheated steam which is treated as an ideal gas of relative molecular mass 18 (relative molecular mass of dry air is 29).

Dry air: the air content of moist air.
Moist air: a mixture of dry air and superheated steam.
Specific humidity (ω) = $\dfrac{\text{mass of steam}}{\text{mass of dry air}}$

$$\omega = \frac{m_s}{m_a} = 0.622\left(\frac{p_s}{p - p_s}\right) \tag{3.6}$$

Relative humidity (ϕ) is also used, where

$$\omega = 0.622\left(\frac{p_g}{p - p_s}\right)\phi \tag{3.7}$$

where p_s is the partial pressure of the steam, p is the mixture total pressure and p_g is the saturation pressure at the air temperature.

Dew point: if an air–steam mixture is cooled at constant pressure the partial pressure of the steam remains constant and, as the cooling proceeds, the temperature will reach the saturation value corresponding to this partial pressure. At this temperature condensing will occur. The mixture has reached the *dew point* (T_d).

Dry bulb temperature: the temperature of the moist air (T_{db}).

Wet bulb temperature: the temperature of the moist air (T_{wb}) recorded on a thermometer whose bulb is enclosed in a water-moistened wick. The difference between T_{db} and T_{wb} can be shown to be dependent on the humidity.

Specific enthalpy and specific volume: these properties are calculated per *unit mass of dry air*; for example, $h = (h_a + \omega\, h_s)$ kJ/kg dry air.

3.9 Properties of reactive mixtures

A common chemical reaction in engineering is the combustion of hydrocarbon fuels in air. In simple combustion problems the sole interest is the energy released by combustion. To solve such problems *calorific value* is defined as the amount of energy released when unit mass of fuel is completely burned under specified experimental conditions. To allow for imperfect combustion in real machinery a combustion efficiency is defined by

$$\eta_{combustion} = \frac{\text{actual energy release}}{\text{ideal energy release}}$$

Calorific values of fuels are readily available data.

This approach will not allow the solution of problems involving chemistry which would require the constituents of the products of combustion to be determined in order to obtain properties of this product mixture. It is not necessary to use chemistry provided suitable specific heat capacities of the products of combustion can be estimated. Data are available for many fuels.

3.10 Interpolation

When using the skeleton tables provided for programs a Lagrangian interpolation[2] method is used. This interpolation is based on the formation of an nth order polynomial from $(n + 1)$ points, which need not be evenly spaced.

3.11 Heat transfer data

Further tables of properties for the solution of heat transfer problems are included in the relevant section.

3.12 Data sources

With the exception of the steam properties, which have been taken from *The Mechanical Engineer's Reference Book*, Butterworths, London (1973), property data has been obtained from Haywood, R.W., *Thermodynamic Tables in S.I. (Metric) Units,* Cambridge University Press (1968). These tables will be found useful for obtaining data for the unprogrammed problems involving substances other than those for which tabulations have been provided.

3.13 References

1. *1967 Steam Tables,* Engineering Research Association, London (1967).
2. Broughton, R.L., and Ramsay, A.R.D., *Numerical Methods,* Bell and Hyman, London (1980).

WORKED EXAMPLES

Example 3.1 Steam data for interpolation

Lines 1010 to 1100 for saturated liquid; lines 1110 to 1200 for dry saturated vapour; lines 1210 to 2200 for superheated vapour.

Pressure range: 0.004 MN/m^2 to 5 MN/m^2.

Each line contains values of: pressure (P) in MN/m^2; temperature (T) in °C; specific enthalpy (H) in kJ/kg; specific entropy (S) in kJ/kg K; specific internal energy (U) in kJ/kg; and specific volume (V) in m^3/kg*

```
            SATURATED LIQUID (N=0 TO 9)

1010  DATA 0.004,29,121.4,0.423,121.4,0.001
1020  DATA 0.01,45.8,191.8,0.649,191.8,0.00101
1030  DATA 0.02,60.1,251.5,0.832,251.4,0.00102
1040  DATA 0.05,81.3,340.6,1.091,340.5,0.00103
1050  DATA 0.1,99.6,417.5,1.303,417.4,0.00104
1060  DATA 0.2,120.2,504.7,1.53,504.5,0.00106
1070  DATA 0.5,151.8,640.1,1.86,639.6,0.00109
1080  DATA 1,179.9,762.6,2.138,761.5,0.00113
1090  DATA 2,212.4,908.6,2.447,906.2,0.00118
1100  DATA 5,263.9,1154.5,2.921,1148,0.00129

            DRY SATURATED VAPOUR (N=10 TO 19)

1110  DATA 0.004,29,2554.5,8.476,2415.3,34.84
1120  DATA 0.01,45.8,2584.8,8.151,2438,14.66
1130  DATA 0.02,60.1,2609.9,7.909,2456.9,7.651
1140  DATA 0.05,81.3,2646,7.595,2484,3.24
1150  DATA 0.1,99.6,2675.4,7.36,2506.1,1.694
1160  DATA 0.2,120.2,2706.3,7.127,2529.2,0.885
1170  DATA 0.5,151.8,2747.5,6.819,2560.2,0.375
1180  DATA 1,179.9,2776.2,6.583,2581.9,0.1943
1190  DATA 2,212.4,2797.2,6.337,2598.2,0.0995
1200  DATA 5,263.9,2794.2,5.974,2597,0.0394

            SUPERHEATED STEAM AT 0.004MN/M↑2 (N=20 TO 29)

1210  DATA 0.004,29,2555,8.476,2415,34.84
1220  DATA 0.004,40,2575,8.453,2431,36.1
1230  DATA 0.004,50,2594,8.602,2445,37.17
1240  DATA 0.004,60,2613,8.659,2459,38.46
1250  DATA 0.004,70,2632,8.715,2474,39.53
1260  DATA 0.004,80,2650,8.769,2487,40.65
1270  DATA 0.004,90,2669,8.822,2502,41.84
1280  DATA 0.004,100,2688,8.873,2516,43.1
1290  DATA 0.004,110,2707,8.923,2530,44.25
1300  DATA 0.004,120,2726,8.972,2545,45.25

            SUPERHEATED STEAM AT 0.01 MN/M↑2 (N=30 TO 39)

1310  DATA 0.01,45.8,2585,8.151,2438,14.66
1320  DATA 0.01,75,2640,8.317,2480,16.03
1330  DATA 0.01,100,2688,8.449,2517,17.18
1340  DATA 0.01,125,2735,8.572,2552,18.35
1350  DATA 0.01,150,2783,8.689,2588,19.53
1360  DATA 0.01,175,2831,8.799,2625,20.66
1370  DATA 0.01,200,2880,8.905,2661,21.83
1380  DATA 0.01,225,2928,9.005,2699,22.99
1390  DATA 0.01,250,2977,9.101,2736,24.15
1400  DATA 0.01,275,3027,9.193,2774,25.32
```

*Capital letters used as symbols here, as the computer programs call these properties with capital letters.

```
            SUPERHEATED STEAM AT 0.02 MN/M↑2 (N=40 TO 49)

1410  DATA 0.02,60.1,2610,7.909,2457,7.651
1420  DATA 0.02,90,2667,8.074,2500,8.354
1430  DATA 0.02,110,2706,8.177,2530,8.818
1440  DATA 0.02,140,2763,8.322,2573,9.515
1450  DATA 0.02,170,2821,8.457,2617,10.21
1460  DATA 0.02,200,2879,8.584,2661,10.91
1470  DATA 0.02,230,2938,8.704,2706,11.6
1480  DATA 0.02,260,2997,8.818,2751,12.3
1490  DATA 0.02,290,3056,8.927,2796,12.99
1500  DATA 0.02,310,3096,8.997,2827,13.46

            SUPERHEATED STEAM AT 0.05 MN/M↑2 (N=50 TO 59)

1510  DATA 0.05,81.3,2646,7.595,2484,3.236
1520  DATA 0.05,125,2731,7.822,2549,3.65
1530  DATA 0.05,150,2780,7.941,2586,3.891
1540  DATA 0.05,200,2878,8.159,2660,4.348
1550  DATA 0.05,250,2976,8.356,2735,4.831
1560  DATA 0.05,300,3076,8.538,2811,5.291
1570  DATA 0.05,350,3177,8.707,2889,5.747
1580  DATA 0.05,400,3279,8.865,2969,6.211
1590  DATA 0.05,425,3331,8.94,3009,6.452
1600  DATA 0.05,450,3383,9.014,3049,6.667

            SUPERHEATED STEAM AT 0.1 MN/M↑2 (N=60 TO 69)

1610  DATA 0.1,99.6,2675,7.36,2506,1.695
1620  DATA 0.1,150,2776,7.614,2583,1.938
1630  DATA 0.1,200,2875,7.835,2658,2.174
1640  DATA 0.1,250,2975,8.034,2734,2.404
1650  DATA 0.1,300,3075,8.217,2811,2.639
1660  DATA 0.1,350,3176,8.386,2889,2.874
1670  DATA 0.1,400,3278,8.544,2968,3.106
1680  DATA 0.1,450,3382,8.6933,3049,3.333
1690  DATA 0.1,500,3488,8.835,3132,3.571
1700  DATA 0.1,550,3596,8.97,3216,3.802

            SUPERHEATED STEAM AT 0.2 MN/M↑2 (N=70 TO 79)

1710  DATA 0.2,120.2,2706,7.127,2529,0.885
1720  DATA 0.2,200,2871,7.507,2655,1.08
1730  DATA 0.2,250,2971,7.71,2731,1.199
1740  DATA 0.2,300,3072,7.894,2809,1.316
1750  DATA 0.2,350,3174,8.064,2887,1.433
1760  DATA 0.2,400,3277,8.223,2967,1.55
1770  DATA 0.2,450,3381,8.372,3048,1.667
1780  DATA 0.2,500,3487,8.514,3131,1.783
1790  DATA 0.2,550,3595,8.649,3216,1.898
1800  DATA 0.2,600,3704,8.778,3301,2.012

            SUPERHEATED STEAM AT 0.5 MN/M↑2 (N=80 TO 89)

1810  DATA 0.5,151.8,2748,6.819,2560,0.375
1820  DATA 0.5,200,2855,7.059,2643,0.425
1830  DATA 0.5,250,2961,7.272,2724,0.4744
1840  DATA 0.5,300,3065,7.461,2803,0.5225
1850  DATA 0.5,350,3168,7.634,2883,0.5701
1860  DATA 0.5,400,3272,7.795,2964,0.6173
1870  DATA 0.5,450,3377,7.945,3045,0.664
1880  DATA 0.5,500,3484,8.088,3128,0.7107
1890  DATA 0.5,550,3592,8.223,3213,0.7576
1900  DATA 0.5,600,3702,8.353,3300,0.8039

            SUPERHEATED STEAM AT 1.0 MN/M↑2(N=90 TO 99)

1910  DATA 1,179.9,2776,6.583,2582,0.1943
1920  DATA 1,225,2886,6.815,2667,0.2196
1930  DATA 1,250,2943,6.926,2710,0.2328
```

```
1940  DATA 1,300,3052,7.125,2794,0.258
1950  DATA 1,350,3159,7.303,2876,0.2824
1960  DATA 1,400,3264,7.467,2958,0.3065
1970  DATA 1,450,3371,7.619,3041,0.3303
1980  DATA 1,500,3478,7.763,3124,0.354
1990  DATA 1,550,3587,7.899,3210,0.3775
2000  DATA 1,600,3697,8.029,3296,0.401

          SUPERHEATED STEAM AT 2.0 MN/M↑2(N=100 TO 109)

2010  DATA 2,212.4,2797,6.337,2598,0.0995
2020  DATA 2,225,2834,6.412,2627,0.1037
2030  DATA 2,250,2902,6.545,2679,0.1115
2040  DATA 2,300,3025,6.77,2774,0.1255
2050  DATA 2,350,3139,6.96,2862,0.1385
2060  DATA 2,400,3240,7.13,2946,0.1511
2070  DATA 2,450,3358,7.286,3031,0.1634
2080  DATA 2,500,3467,7.432,3116,0.1754
2090  DATA 2,550,3578,7.571,3202,0.1876
2100  DATA 2,600,3689,7.702,3290,0.1996

          SUPERHEATED STEAM AT 5 MN/M↑2 (N=110 TO 119)

2110  DATA 5,263.9,2794,5.974,2597,0.0394
2120  DATA 5,290,2892,6.152,2673,0.0438
2130  DATA 5,320,2987,6.316,2747,0.0481
2140  DATA 5,360,3098,6.497,2832,0.0532
2150  DATA 5,400,3198,6.651,2909,0.0578
2160  DATA 5,440,3294,6.789,2983,0.0622
2170  DATA 5,480,3387,6.916,3055,0.0664
2180  DATA 5,520,3480,7.036,3127,0.0706
2190  DATA 5,560,3572,7.149,3199,0.0746
2200  DATA 5,600,3665,7.258,3272,0.0786
```

Steam table warning

Care is needed for interpolations in the saturated liquid and dry saturated vapour tables that involve a large pressure range. It may be necessary to restrict the range of N values used for interpolation to get sensible results. Care is also needed when volume is used as the interpolation variable since volume decreases as pressure increases and it may be preferable to use density. It should be noted that between 2 MN/m^2 and 5 MN/m^2 the dry saturated vapour value of specific enthalpy and internal energy passes through a maximum. This may lead to error and if it is intended to use the skeleton table for repeated work in this region it would be advisable to add extra data at pressures greater than 5 MN/m^2.

Example 3.2 Refrigerant 12 data for interpolation

Lines 3010 to 3200 for saturated liquid; lines 3210 to 3400 for dry saturated vapour; lines 3410 to 4000 for superheated vapour.

Temperature range: –60°C to +50°C.

Each line contains values of: temperature (T) in °C; pressure (P) in MN/m^2; specific volume (V) in m^3/kg; specific enthalpy (H) in kJ/kg; and specific enthalpy (S) in kJ/kg K.

```
          SATURATED LIQUID (N=0 TO 19)

3010  DATA -60,0.0226,0.00065,-17.5,-0.0783
3020  DATA -50,0.0392,0.00066,-8.8,-0.0384
3030  DATA -40,0.0641,0.00066,0,0
3040  DATA -35,0.0806,0.00067,4.[illegible],0.0187
3050  DATA -30,0.1003,0.00067,8.9,0.0371
3060  DATA -25,0.1236,0.00068,13.3,0.0552
3070  DATA -20,0.1508,0.00069,17.8,0.0731
3080  DATA -15,0.1825,0.00069,22.3,0.0906
3090  DATA -10,0.219,0.0007,26.9,0.108
3100  DATA -5,0.261,0.00071,31.4,0.1251
3110  DATA 0,0.308,0.00072,36.1,0.142
3120  DATA 5,0.362,0.00072,40.7,0.1587
3130  DATA 10,0.423,0.00073,45.4,0.1752
3140  DATA 15,0.491,0.00074,50.1,0.1915
3150  DATA 20,0.567,0.00075,54.9,0.2078
3160  DATA 25,0.651,0.00076,59.7,0.2239
3170  DATA 30,0.745,0.00077,64.6,0.2399
3180  DATA 35,0.847,0.00079,69.5,0.2559
3190  DATA 40,0.96,0.0008,74.6,0.2718
3200  DATA 50,1.219,0.00083,84.9,0.3037

          DRY SATURATED VAPOUR (N=20 TO 39)

3210  DATA -60,0.0226,0.6379,160.3,0.7558
3220  DATA -50,0.0392,0.3831,165,0.7401
3230  DATA -40,0.0641,0.2421,169.6,0.7274
3240  DATA -35,0.0806,0.1955,171.9,0.722
3250  DATA -30,0.1003,0.1595,174.2,0.7171
3260  DATA -25,0.1236,0.1313,176.5,0.7127
3270  DATA -20,0.1508,0.1089,178.7,0.7088
3280  DATA -15,0.1825,0.0911,181,0.7052
3290  DATA -10,0.219,0.0767,183.2,0.702
3300  DATA -5,0.261,0.065,185.4,0.6991
3310  DATA 0,0.308,0.0554,187.5,0.6966
3320  DATA 5,0.362,0.0475,189.7,0.6942
3330  DATA 10,0.423,0.0409,191.7,0.6921
3340  DATA 15,0.491,0.0354,193.8,0.6902
3350  DATA 20,0.567,0.0308,195.8,0.6885
3360  DATA 25,0.651,0.0269,197.7,0.6869
3370  DATA 30,0.745,0.0235,199.6,0.6854
3380  DATA 35,0.847,0.0206,201.5,0.6839
3390  DATA 40,0.96,0.0182,203.2,0.6825
3400  DATA 50,1.219,0.0142,206.5,0.6797

          SUPERHEATED REFRIGERANT 12(N=40 TO 99)

3410  DATA -60,0.0226,0.638,160.3,0.7588
3420  DATA -40,0.0226,0.689,170.9,0.803
3430  DATA -20,0.0266,0.745,181.7,0.8488


3440  DATA -50,0.0392,0.383,165,0.7401
3450  DATA -30,0.0392,0.418,175.9,0.786[illegible]
3460  DATA -10,0.0392,0.452,187,0.8319


3470  DATA -40,0.0641,0.242,169.6,0.7274
3480  DATA -20,0.0641,0.264,180.8,0.7737
3490  DATA 0,0.0641,0.283,192.4,0.8178


3500  DATA -35,0.0806,0.196,171.7,0.722
3510  DATA -15,0.0806,0.214,183.3,0.7681
3520  DATA 5,0.0806,0.23,195.1,0.812


3530  DATA -30,0.1003,0.16,174.2,0.7171
3540  DATA -10,0.1003,0.173,185.8,0.7631
3550  DATA 10,0.1003,0.188,197.8,0.8068
```

```
3560 DATA -25,0.1236,0.131,176.5,0.7127
3570 DATA -5,0.1236,0.142,188.3,0.7586
3580 DATA 15,0.1236,0.155,200.4,0.8021

3590 DATA -20,0.1508,0.109,178.7,0.7088
3600 DATA 0,0.1508,0.119,190.8,0.7546
3610 DATA 20,0.1508,0.125,203.1,0.7979

3620 DATA -15,0.1825,0.091,181,0.7052
3630 DATA 5,0.1825,0.1,193.2,0.751
3640 DATA 25,0.1825,0.106,205.7,0.7942

3650 DATA -10,0.219,0.077,183.2,0.702
3660 DATA 10,0.219,0.083,195.7,0.7477
3670 DATA 30,0.219,0.09,208.3,0.7909

3680 DATA -5,0.261,0.065,185.4,0.6991
3690 DATA 15,0.261,0.069,198.1,0.7449
3700 DATA 35,0.261,0.076,210.9,0.7879

3710 DATA 0,0.308,0.055,187.5,0.6966
3720 DATA 20,0.308,0.057,200.5,0.7423
3730 DATA 40,0.308,0.064,213.5,0.7853

3740 DATA 5,0.362,0.048,189.7,0.6942
3750 DATA 25,0.362,0.049,202.9,0.7401
3760 DATA 45,0.362,0.055,216.1,0.783

3770 DATA 10,0.423,0.041,191.7,0.6921
3780 DATA 30,0.423,0.042,205.2,0.7381
3790 DATA 50,0.423,0.047,218.6,0.781

3800 DATA 15,0.491,0.035,193.8,0.6902
3810 DATA 35,0.491,0.036,207.5,0.7363
3820 DATA 55,0.491,0.039,221.2,0.7792

3830 DATA 20,0.567,0.031,195.8,0.6885
3840 DATA 40,0.567,0.032,209.8,0.7348
3850 DATA 60,0.567,0.035,223.7,0.7777

3860 DATA 25,0.651,0.027,197.7,0.6869
3870 DATA 45,0.651,0.028,212.1,0.7334
3880 DATA 65,0.651,0.031,226.1,0.7763

3890 DATA 30,0.745,0.024,199.6,0.6854
3900 DATA 50,0.745,0.025,214.3,0.7321
3910 DATA 70,0.745,0.028,228.6,0.7751

3920 DATA 35,0.847,0.021,201.5,0.6839
3930 DATA 55,0.847,0.022,216.4,0.731
3940 DATA 75,0.847,0.025,231,0.7741

3950 DATA 40,0.96,0.018,203.2,0.6825
3960 DATA 60,0.96,0.019,218.5,0.73
3970 DATA 80,0.96,0.022,233.4,0.7732

3980 DATA 50,1.219,0.014,206.5,0.6797
3990 DATA 70,1.219,0.015,222.6,0.7282
4000 DATA 90,1.219,0.017,238,0.7718
```

Data note

When using these data blocks (which will be on tape, disc or EPROM), it is necessary to bring them into the program. Intelligent selection of the amount taken in can reduce the computer storage needed. In the program listings shown here all the data for a working substance is read in and then selected by the values of N in the interpolation routine. In the particular computer used the 120 lines of steam data are labelled N = 0 to 119 in the store. Thus, for example, to select only the 2 MN/m^2 superheat data for use, N is set from 100 to 109. Similarly, for the refrigeration data the 100 lines are labelled N = 0 to 99 in the store.

```
DEMONSTRATION INTERPOLATION ROUTINE

LIST
 10  PRINT "LAGRANGIAN INTERPOLATION FOR H GIVEN T"
 20  PRINT
 30  DIM T[5],H[5]
 40  FOR N=0 TO 4
 50   READ T[N],H[N]
 60  NEXT N
 70  PRINT "WHAT IS THE VALUE OF T FOR INTERPOLATION?"
 80  INPUT A
 90  B=0
 100  FOR J=0 TO 4
 110   X=1
 120   FOR N=0 TO 4
 130    IF N=J THEN GOTO 150
 140    X=X*(A-T[N])/(T[J]-T[N])
 150   NEXT N
 160   B=B+X*H[J]
 170  NEXT J
 180  PRINT " AT THE GIVEN VALUE OF T,Y=";B
 190  REM GIVE DATA FOR INTERPOLATION
 200  DATA 1,1
 210  DATA 2,2
 220  DATA 3,3
 230  DATA 4,4
 240  DATA 6,6
 250  END

RUN
LAGRANGIAN INTERPOLATION FOR H GIVEN T

WHAT IS THE VALUE OF T FOR INTERPOLATION?
? 5
 AT THE GIVEN VALUE OF T,Y= 5

STOP AT 250
```

Program notes

(1) Lines 40 to 60 read in data from lines 200 to 240 to store in the blocks of 5 spaces allocated by line 30.
(2) Lagrangian interpolation theory (section 3.10) results in a relation for $(n + 1)$ points $(x_0, y_0), (x_1, y_1), \ldots\ldots (x_n, y_n)$

$$y(x) = \sum_{i=0}^{i=n} \phi_i(x)\, y_i$$

where

$$\phi_i(x) = \frac{(x - x_0)(x - x_1)\ldots\ldots(x - x_{i-1})(x - x_{i+1})\ldots\ldots(x - x_n)}{(x_i - x_0)(x_i - x_1)\ldots\ldots(x_i - x_{i+1})(x_1 - x_{i+1})\ldots\ldots(x_i - x_n)}$$

$$= \prod_{k=0}^{n} \frac{(x - x_k)}{(x_i - x_k)} \quad [k \neq i]$$

The interpolation routine is from line 90 to line 170 and this block will need to be adapted to suit individual problems.

(3) This example is trivial but the routine can be tested on other blocks of data.

Example 3.3 Steam table interpolation

Determine the specific entropy of superheated steam at 0.01 MN/m^2, 160°C.

```
LIST
 10  PRINT "EXAMPLE 3.3
 20  PRINT
 30  DIM P[120],T[120],H[120],S[120],U[120],V[120]
 40  FOR N=0 TO 119
 50   READ P[N],T[N],H[N],S[N],U[N],V[N]
 60  NEXT N
 70  B=0
 80  PRINT "TEMPERATURE?"
 90  INPUT A
 100  FOR J=30 TO 39
 110   X=1
 120   FOR N=30 TO 39
 130    IF N=J THEN GOTO 150
 140    X=X*(A-T[N])/(T[J]-T[N])
 150   NEXT N
 160   B=B+X*S[J]
 170  NEXT J
 180  PRINT "AT 0.01 MN/M↑2,T=160 C,S=";B;" KJ/KG K"
 190  END


RUN
EXAMPLE 3.3

TEMPERATURE?
? 160
AT 0.01 MN/M↑2,T=160 C,S= 8.73366298 KJ/KG K

STOP AT 190
```

Program notes

(1) In lines 40 to 60 the steam tables are read into the space allocated in line 30.

(2) Interpolation takes place in lines 100 to 170 in the 0.01 MN/m^2 pressure block. This data block is selected by the values assigned to

J and N. As an alternative to this, the whole steam table superheat data could be scanned to select the appropriate block. If the pressure was not one of those in the data tables, the pressure blocks which bracket the required value may be located visually or by an interpolation routine. Values of specific entropy at 160°C can then be determined as in the program and these two values may be linearly interpolated to determine the specific entropy at the required pressure and temperature. This is shown in the superheat part of example 3.5. It can be seen that careful program construction will reduce the amount of interpolation required.

Example 3.4 Refrigerant table interpolation

Determine the specific enthalpy of refrigerant 12 at a pressure and temperature of 0.62 MN/m^2, 55°C.

```
LIST
 10  PRINT "EXAMPLE 3.4"
 20  DIM T[100],P[100],V[100],H[100],S[100]
 30  FOR N=0 TO 99
 40   READ T[N],P[N],V[N],H[N],S[N]
 50  NEXT N
 60  B=0
 70  PRINT "TEMPERATURE?"
 80  INPUT A
 90  FOR J=82 TO 84
 100   X=1
 110   FOR N=82 TO 84
 120    IF N=J THEN GOTO 140
 130    X=X*(A-T[N])/(T[J]-T[N])
 140   NEXT N
 150   B=B+X*H[J]
 160  NEXT J
 170  H1=B
 180  C=0
 190  FOR J=85 TO 87
 200   Z=1
 210   FOR N=85 TO 87
 220    IF N=J THEN GOTO 240
 230    Z=Z*(A-T[N])/(T[J]-T[N])
 240   NEXT N
 250   C=C+Z*H[J]
 260  NEXT J
 270  H2=C
 280  H=H1+(H2-H1)*(0.62-P[82])/(P[85]-P[82])
 290  PRINT "SPECIFIC ENTHALPY AT 0.62MN/M↑2=";H;"KJ/KG"
 300  PRINT "A BETTER FORMAT FOR RESULTS MIGHT BE TO CUT OFF"
 305  PRINT "SOME OF THE DECIMAL PLACES"
 310  H=1000*H
 320  H=INP[H]:: H=H/1000
 330  PRINT "SPECIFIC ENTHALPY AT 0.62 MN/M↑2";H;" KJ/KG"
 340  END
 3009  PRINT
 3010  DATA -60,0.0226,0.00065,-17.5,-0.0783
 3020  DATA -50,0.0392,0.00066,-8.8,-0.0384
 3030  DATA -40,0.0641,0.00066,0,0


RUN
EXAMPLE 3.4
TEMPERATURE?
? 55
```

```
SPECIFIC ENTHALPY AT 0.62MN/M↑2= 219.550186KJ/KG
A BETTER FORMAT FOR RESULTS MIGHT BE TO CUT OFF
SOME OF THE DECIMAL PLACES
SPECIFIC ENTHALPY AT 0.62 MN/M↑2 219.55 KJ/KG

STOP AT 340
```

Program notes

(1) By inspection of the skeleton tables it is seen that the refrigerant is superheated.
(2) Lines 30 to 50 read in data to store in the space allocated by line 20.
(3) Lines 90 to 270 interpolate in the pressure blocks with pressures less than and greater than 0.62 MN/m^2 to find specific enthalpy at 55°C.
(4) Lines 280 and 290 interpolate between these two specific enthalpy values to obtain the data at the intermediate pressure.
(5) In line 320 the double colon has been used to allow more than one statement to appear on one line. This technique has only rarely been used in the remaining programs but may be incorporated if required and available.
(6) In lines 300 to 330 the large number of decimal places has been reduced by the use of the INP function. This method has not been used in other programs but could be incorporated if required.
(7) A more elegant method of reducing the number of figures displayed which includes a rounding up factor is to use the single statement

H=(INP(H*1000+0.5))/1000

Example 3.5 Steam table interpolation

Determine the specific enthalpy of 1 kg of steam occupying a volume of (a) 0.06 m^3 at 4 MN/m^2, and (b) 0.04 m^3 at 4 MN/m^2.

The necessary reading for this example is in section 3.3.

```
LIST
10  PRINT "EXAMPLE  .  "
20  PRINT
30  DIM P[120],T[120],H[120],S[120],U[120],V[120]
40  FOR N=0 TO 119
50    READ P[N],T[N].H[N],S[N],U[N],V[N]
60  NEXT N
70  PRINT "FIND VG AT THE REQUIRED PRESSURE"
80  DIM RO[120]
90  FOR N=0 TO 119
100   RO[N]=1/V[N]
110  NEXT N
120  ROG=0
130  HG=0
140  PRINT "PRESSURE"
```

```
15Ø  INPUT A
16Ø  FOR J=17 TO 19
17Ø   X=1
18Ø   FOR N=17 TO 19
19Ø    IF N=J THEN GOTO 21Ø
2ØØ    X=X*(A-P[N])/(P[J]-P[N])
21Ø   NEXT N
22Ø   ROG=ROG+X*RO[J]
23Ø   HG=HG+X*H[J]
24Ø  NEXT J
25Ø  VG=1/ROG
26Ø  PRINT "VG=";VG;"M↑3/KG"
27Ø  PRINT "HG=";HG;"KJ/KG"
28Ø  PRINT "INPUT VOLUME"
29Ø  INPUT D
3ØØ  PRINT "IS VG GREATER THAN INPUT VOLUME?YES TYPE 1,NØ TYPE 2"
31Ø  INPUT K
32Ø  IF K=2 THEN GOTO 52Ø
33Ø  PRINT "WET STEAM"
34Ø  VF=Ø
35Ø  HF=Ø
36Ø  FOR J=7 TO 9
37Ø   B=1
38Ø   FOR N=7 TO 9
39Ø    IF N=J THEN GOTO 41Ø
4ØØ    B=B*(A-P[N])/(P[J]-P[N])
41Ø   NEXT N
42Ø   VF=VF+B*V[J]
43Ø   HF=HF+B*H[J]
44Ø  NEXT J
45Ø  PRINT "VF=";VF;"M↑3/KG"
46Ø  PRINT "HF=";HF;"KJ/KG"
47Ø  DF=(D-VF)/(VG-VF)
48Ø  PRINT "DRYNESS FRACTION=";DF
49Ø  HW=HF+DF*(HG-HF)
5ØØ  PRINT "SPECIFIC ENTHALPY=";HW;"KJ/KG"
51Ø  GOTO 93Ø
52Ø  PRINT "SUPERHEATED STEAM"
53Ø  PRINT "INTERPOLATE AT P>REQUIRED PRESSURE"
54Ø  C=Ø
55Ø  FOR J=11Ø TO 119
56Ø   Y=1
57Ø   FOR N=11Ø TO 119
58Ø    IF N=J THEN GOTO 6ØØ
59Ø    Y=Y*(D-V[N])/(V[J]-V[N])
6ØØ   NEXT N
61Ø   C=C+Y*H[J]
62Ø  NEXT J
63Ø  H1=C
64Ø  PRINT "INTERPOLATE AT P<REQUIRED PRESSURE"
65Ø  REM IN THIS CASE P<REQUIRED PRESSURE IS LINE 1198(N=18) FOR G VALUES
66Ø  N=18
67Ø  VG=V[N]
68Ø  HG=H[N]
685  PRINT "VG=";VG"M↑3/KG".
69Ø  PRINT "IS V GREATER THAN VG?YES TYPE 1,NO TYPE 2"
7ØØ  INPUT L
71Ø  IF L=2 THEN GOTO 83Ø
72Ø  E=Ø
73Ø  FOR J=1ØØ TO 1Ø9
74Ø   Z=1
75Ø   FOR N=1ØØ TO 1Ø9
76Ø    IF N=J THEN GOTO 78Ø
77Ø    Z=Z*(D-V[N])/(V[J]-V[N])
78Ø   NEXT N
79Ø   E=E+Z*H[J]
8ØØ  NEXT J
81Ø  H2=E
82Ø  GOTO 91Ø
83Ø  PRINT "SUPERHEATED STEAM BUT LOWER PRESSURE IS WET"
84Ø  REM IN THIS CASE  P<REQUIRED VALUE IS LINE 1Ø9Ø(N=8) FOR F VALUES
```

```
850  N=8
860  VF=V[N]
870  HF=H[N]
880  DF=(D-VF)/(VG-VF)
890  PRINT "DRYNESS FRACTION";DF
900  H2=HF+DF*(HG-HF)
910  HS=H2+(H1-H2)*(A-P[8])/(P[9]-P[8])
920  PRINT "SPECIFIC ENTHALPY=";HS;"KJ/KG"
930  END


RUN
EXAMPLE 3.5

FIND VG AT THE REQUIRED PRESSURE
PRESSURE
? 4
VG= 0.0495853023M↑3/KG
HG= 2806.2KJ/KG
INPUT VOLUME
? .06
IS VG GREATER THAN INPUT VOLUME?YES TYPE 1,NO TYPE 2
? 2
SUPERHEATED STEAM
INTERPOLATE AT P>REQUIRED PRESSURE
INTERPOLATE AT P<REQUIRED PRESSURE
VG= 0.0995M↑3/KG
IS V GREATER THAN VG?YES TYPE 1,NO TYPE 2
? 2
SUPERHEATED STEAM BUT LOWER PRESSURE IS WET
DRYNESS FRACTION 0.598250610
SPECIFIC ENTHALPY= 2843.36492KJ/KG

STOP AT 930


RUN
EXAMPLE 3.5

FIND VG AT THE REQUIRED PRESSURE
PRESSURE
? 4
VG= 0.0495853023M↑3/KG
HG= 2806.2KJ/KG
INPUT VOLUME
? .04
IS VG GREATER THAN INPUT VOLUME?YES TYPE 1,NO TYPE 2
? 1
WET STEAM
VF= 0.00126M↑3/KG
HF= 1104.55KJ/KG
DRYNESS FRACTION= 0.801650444
SPECIFIC ENTHALPY= 2468.67848KJ/KG

STOP AT 930
```

Program nomenclature

RO	density
ROG	saturated steam density
HG	saturated steam specific enthalpy
VG	saturated steam specific volume
VF	saturated liquid specific volume
HF	saturated liquid specific enthalpy

DF dryness fraction
HW wet steam specific enthalpy
HS superheated steam specific enthalpy

Program notes

(1) This program is more complex and requires a careful plan:
(i) find v_g at 4 MN/m^2;
(ii) compare v with v_g to find if wet or superheated (it is unlikely that v will be exactly equal to v_g);
(iii) if wet, find x and use $h = h_f + x\, h_{fg}$;
(iv) if superheated interpolate between pressure blocks bracketing required pressure, but there are two possible cases (Figure 3.3): (a) both pressure blocks are superheated, or (b) the lower pressure block is wet.
(v) Figure 3.3 shows by means of the arrows the data routes to determine the specific enthalpy for cases (iii) and (iv)(a) or (iv)(b).

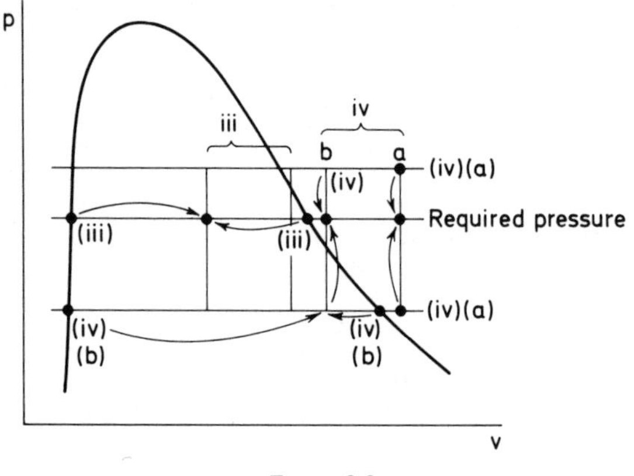

Figure 3.3

Thus in the lower pressure block (of the two bracketing the required pressure) a further test for wet or superheated must be made so that it can be decided whether to use wet or superheated interpolation. This interpolation may be arranged to coincide with the existing superheated or wet interpolation if required.
(2) Lines 30 to 60 read in data.
(3) Lines 80 to 100 change v to ρ because the pressure–volume saturated liquid relation is hyperbolic and unsuited to interpolation. The density–pressure relation is more satisfactory.
(4) Lines 130 to 270 find v_g and h_g. The J and N range is restricted since interpolation over a large pressure range is not suitable.

(5) Lines 300 to 320 decide 'wet or superheated'.
(6) Lines 330 to 510 find wet steam enthalpy. It can be seen that the J and N range is restricted to avoid interpolation over a large pressure range.
(7) Lines 520 to 630 interpolate in the superheat pressure greater than 4 MN/m^2.
(8) Lines 640 to 820 interpolate in the superheat pressure less than 4 MN/m^2 (not used in this example).
(9) Lines 830 to 900 interpolate in the wet steam zone. (Case (iv)(b) above.)

Example 3.6 Gas properties

The properties R, C_v, C_p and γ of a perfect gas are related by the equations

$$C_p - C_v = R$$

$$\frac{C_p}{C_v} = \gamma$$

Write a program to obtain the two unknown properties given any two properties. (Note: since $RM = 8.3143$ kJ/kmol K, M is the alternative to R.)

Test the program (a) by finding the values of γ and C_p for air for which $R = 0.287$ kJ/kg K and $C_v = 0.72$ kJ/kg K and (b) by finding the values of R and C_v for nitrogen for which $C_p = 1.041$ kJ/kg K and $\gamma = 1.4$.

```
LIST
 10  PRINT "EXAMPLE 3.6"
 20  PRINT
 30  PRINT "WHAT ARE THE INPUTS"
 40  PRINT "R AND CP,OPTION 1"
 50  PRINT "R AND CV,OPTION 2"
 60  PRINT "R AND GAMMA,OPTION 3"
 70  PRINT "CP AND CV,OPTION  4"
 80  PRINT "CP AND GAMMA,OPTION 5"
 90  PRINT "CV AND GAMMA,OPTION 6"
 100  INPUT X
 110  IF X=1 THEN GOTO 170
 120  IF X=2 THEN GOTO 230
 130  IF X=3 THEN GOTO 290
 140  IF X=4 THEN GOTO 350
 150  IF X=5 THEN GOTO 410
 160  IF X=6 THEN GOTO 470
 170  INPUT R,CP
 180  G=1/(1-(R/CP))
 190  CV=CP-R
 200  PRINT "GAMMA IS";G
 210  PRINT "CV IS";CV;"KJ/KG K"
 220  GOTO 520
 230  INPUT R;CV
 240  G=(R/CV)+1
 250  CP=R+CV
 260  PRINT "GAMMA IS";G
 270  PRINT "CP IS";CP;"KJ/KG K"
 280  GOTO 520
 290  INPUT R;G
 300  CV=R/(G-1)
 310  CP=CV*G
 320  PRINT "CP IS";CP;"KJ/KG K"
```

```
330  PRINT "CV IS";CV;"KJ/KG K"
340  GOTO 520
350  INPUT CP;CV
360  R=CP-CV
370  G=CP/CV
380  PRINT "R IS";R;"KJ/KG K"
390  PRINT "G IS";G
400  GOTO 520
410  INPUT CP;G
420  CV=CP/G
430  R=CP-CV
440  PRINT "R IS";R;"KJ/KG K"
450  PRINT "CV IS";CV;"KJ/KG K"
460  GOTO 520
470  INPUT CV;G
480  CP=CV*G
490  R=CP-CV
500  PRINT "R IS";R;"KJ/KG K"
510  PRINT "CP IS";CP;"KJ/KG K"
520  END
```

```
PROPERTIES OF AIR (PERFECT GAS)

RUN
EXAMPLE  .

WHAT ARE THE INPUTS
R AND CP,OPTION 1
R AND CV,OPTION 2
R AND GAMMA,OPTION 3
CP AND CV,OPTION  4
CP AND GAMMA,OPTION 5
CV AND GAMMA,OPTION 6
? 2
? 0.2870.72
GAMMA IS 1.39861111
CP IS 1.007KJ/KG K

STOP AT 520
```

```
PROPERTIES OFNITROGEN(PERFECT GAS)

RUN
EXAMPLE  .

WHAT ARE THE INPUTS
R AND CP,OPTION 1
R AND CV,OPTION 2
R AND GAMMA,OPTION 3
CP AND CV,OPTION  4
CP AND GAMMA,OPTION 5
CV AND GAMMA,OPTION 6
? 5
? 1.041.4
R IS 0.297142857KJ/KG K
CV IS 0.742857143KJ/KG K

STOP AT 520
```

Program nomenclature

R specific gas constant
CP specific heat at constant pressure
CV specific heat at constant volume
G gamma

Program note

This program contains six mini-programs to calculate the missing data. In lines 40–160 the appropriate mini-program is chosen.

Example 3.7 Psychrometric properties

Write a program to determine the properties of atmospheric air for an entry of dry bulb temperature and relative humidity. Test the program by finding h, v and ω for air at (a) 30°C dry bulb temperature and 60% relative humidity and (b) 15°C dry bulb temperature and 20% relative humidity. In the first case the dew point temperature should also be determined.

The necessary reading for this example is in section 3.8.

```
LIST
 4010  PRINT "EXAMPLE 3.7"
 4020  PRINT "PSYCHROMETRIC PROPERTIES,INPUT TDB AND PHI,AT 101.325 MN/M↑2"
 4030  DIM T[8],P[8]
 4040  FOR N=0 TO 7
 4050   READ T[N],P[N]
 4060  NEXT N
 4070  REM INTERPOLATE FOR PG AT TDB
 4080  B=0
 4090  PRINT "TDB?"
 4100  INPUT TDB
 4110  FOR J=0 TO 7
 4120   X=1
 4130   FOR N=0 TO 7
 4140    IF N=J THEN GOTO 4160
 4150    X=X*(TDB-T[N])/(T[J]-T[N])
 4160   NEXT N
 4170   B=B+X*P[J]
 4180  NEXT J
 4190  PRINT "PG=";B;"KN/M↑2"
 4200  PRINT "PHI?"
 4210  INPUT PHI
 4220  PS=PHI*B
 4230  PRINT "PS=";PS;"KN/M↑2"
 4240  W=0.622*PS/(101.325-PS)
 4250  H=(1.01*TDB)+W*(2501.6+1.86*TDB)
 4260  V=(0.287*(TDB+273))/(101.325-PS)
 4270  PRINT "ENTHALPY DATUM IS WATER AT 0.01 C"
 4280  PRINT "SPECIFIC ENTHALPY,H=";H;"KJ/KG DRY AIR"
 4290  PRINT "SPECIFIC VOLUME,V=";V;"M↑3/KG DRY AIR"
 4300  PRINT "SPECIFIC HUMIDITY,W=";W
 4305  PRINT "IF PS<0.611 DEW POINT TEMP IS BELOW 0 C AND PROGRAM INVALID'
 4310  PRINT "DO YOU WANT DEW POINT TEMPERATURE?YES,TYPE 1,NO,TYPE 2"
 4320  INPUT D
 4330  IF D=2 THEN GOTO 4450
 4340  PRINT "INTERPOLATE FOR DEW POINT AT PS"
 4350  C=0
 4360  FOR J=0 TO 7
 4370   Z=1
 4380   FOR N=0 TO 7
 4390    IF N=J THEN GOTO 4410
 4400    Z=Z*(PS-P[N])/(P[J]-P[N])
 4410   NEXT N
 4420   C=C+Z*T[J]
 4430  NEXT J
 4440  PRINT "DEW POINT TEMPERATURE=";C;" C"
 4450  END
 4460  DATA 0.01,0.611
 4470  DATA 5,0.872
 4480  DATA 10,1.227
```

```
4490  DATA 15,1.704
4500  DATA 20,2.337
4510  DATA 25,3.166
4520  DATA 30,4.241
4530  DATA 35,5.622


RUN
EXAMPLE  .
PSYCHROMETRIC PROPERTIES,INPUT TDB AND PHI,AT 101.325 MN/M↑2
TDB?
? 30
PG= 4.241KN/M↑2
PHI?
? 0.6
PS= 2.5446KN/M↑2
ENTHALPY DATUM IS WATER AT 0.01 C
SPECIFIC ENTHALPY,H= 71.2767762KJ/KG DRY AIR
SPECIFIC VOLUME,V= 0.8803467M↑3/KG DRY AIR
SPECIFIC HUMIDITY,W= 0.0160228264
IF PS<0.611 DEW POINT TEMP IS BELOW 0 C AND PROGRAM INVALID
DO YOU WANT DEW POINT TEMPERATURE?YES,TYPE 1,NO,TYPE 2
? 1
INTERPOLATE FOR DEW POINT AT PS
DEW POINT TEMPERATURE= 21.3711570 C

STOP AT 4450


RUN
EXAMPLE  .
PSYCHROMETRIC PROPERTIES,INPUT TDB AND PHI,AT 101.325 MN/M↑2
TDB?
? 15
PG= 1.704KN/M↑2
PHI?
? 0.2
PS= 0.3408KN/M↑2
ENTHALPY DATUM IS WATER AT 0.01 C
SPECIFIC ENTHALPY,H= 20.4597152KJ/KG DRY AIR
SPECIFIC VOLUME,V= 0.818504281M↑3/KG DRY AIR
SPECIFIC HUMIDITY,W= 0.00209911650
IF PS<0.611 DEW POINT TEMP IS BELOW 0 C AND PROGRAM INVALID
DO YOU WANT DEW POINT TEMPERATURE?YES,TYPE 1,NO,TYPE 2
? 2

STOP AT 4450
```

Program nomenclature

TDB	dry bulb temperature
PG	partial pressure of saturated steam
PHI	relative humidity
PS	partial pressure of superheated steam
W	specific humidity
H	specific enthalpy
V	specific volume

Program notes

(1) A low-temperature data block is included in lines 4460 to 4530 which gives saturation temperature and pressure.

(2) Specific enthalpy of the mixture is obtained by

$$h = h_a + \omega h_s$$

$$h \doteq C_{p_a} T_{db} + \omega (h_{g0.01} + C_{p_s} T_{db})$$

where $\omega = 0.622 \left(\frac{p_s}{p - p_s}\right)$

with $C_{p_a} = 1.01$ kJ/kg K and $C_{p_s} = 1.86$ kJ/kg K

(3) Specific volume of the mixture is obtained by

$$v = \frac{R_a T}{p_a} \qquad \text{with } R_a = 0.287 \text{ kJ/kg K}$$

(4) An option to calculate dew point temperature is allowed by lines 4310 to 4330. (In air-conditioning calculations dew point temperature is not always required.)
(5) Humidity is determined by measuring dry bulb temperature and wet bulb temperature and then tables or charts based on mass transfer considerations are used. The equations given in problem 3.5 enable this program to be reconstructed for use with input data of T_{db} and T_{wb}.
(6) This program has lines numbered from 4010 to 4530 to enable use as a data subroutine later.

PROBLEMS

(3.1) Determine the specific internal energy of steam at 0.05 MN/m^2: (a) at 225°C, (b) with specific entropy 7.28 kJ/kg K.
(3.2) Determine the specific enthalpy of refrigerant 12 at a temperature of –13°C with specific entropy = 0.12 kJ/kg K.
(3.3) Make a data block for ammonia (refrigerant) and hence determine the specific enthalpy at a temperature of –5°C with specific entropy 1.01 kJ/kg K (store the data block for possible future use).
(3.4) An orifice of diameter D with coefficient of discharge C_D is used to measure the volume flow rate of air ($V = AC_D\sqrt{2gh}$ where A is the orifice area and h is the pressure drop in metres of air). Write a program to give the mass flow rate of air and hence plot (by hand) a calibration curve for the orifice ($\dot{m}$ vs h). Test results are with air temperature 28°C, orifice diameter, 41 mm; barometer, 101 kN/m^2 and $C_D = 0.63$:

h; mm H_2O	110	92	67	36	16	8
Orifice; gauge pressure; mm H_2O	470	432	345	195	95	43

(3.5) Write a program to determine the specific humidity, relative humidity, specific enthalpy and specific volume of atmospheric air based on measurements of dry bulb and wet bulb temperature. Use example 3.7 to assist, and include the dew point option. Test the program with the data of example 3.7.

An *approximate* relation based on mass transfer considerations to determine the partial pressure of the steam (p_s) in the mixture is

$$p_s = \frac{(273 + T_{db})}{(273 + T_{wb})}\left[p_{gw} - \frac{148.19\,(T_{db} - T_{wb})}{h_{fgw}}\right] \text{kN/m}^2$$

where p_{gw} = saturation pressure at T_{wb} in kN/m^2
and h_{fgw} = specific enthalpy of vaporisation at T_{wb}.

(A simple explanation of the wet and dry bulb processes may be found in Simonson, J.R., *Engineering Heat Transfer,* MacMillan, London (1975), pages 155 to 157.)

(3.6) Clapeyron's equation for phase change is

$$\left(\frac{dp}{dT}\right)_{sat} = \frac{h_g - h_f}{T_{sat}\,(v_g - v_f)}$$

(T_{sat} must be in kelvins.) Use this equation to evaluate $(dp/dT)_{sat}$ for steam over the pressure range 0.004 MN/m^2 to 5 MN/m^2. Hand plot the shape of the p–T curve to obtain $(dp/dT)_{sat}$ values for comparison.

(3.7) Further problems in data use can be invented and results checked with tables or charts. Agreement may not be exact as the data blocks provided are sparse. Do not attempt to extrapolate the blocks beyond the pressure and temperature ranges listed.

Chapter 4

Processes and cycles

ESSENTIAL THEORY

4.1 Introduction

A number of ideal reversible processes are used to form ideal cycles equivalent to various real machine cycles. These are considered for non-flow and flow situations and the relevant property changes, work transfers and heat transfers associated with the processes are evaluated and used to determine cycle parameters.

Two power plant cycle parameters of interest to engineers are the specific work transfer (or specific output) and the thermal efficiency. As an alternative to the specific output, specific consumption of working substance may be used. These parameters will not apply to air compression cycles nor to refrigeration, heat pump and air-conditioning cycles.

Specific work transfer (w)

$$= \text{net work/kg working substance} \left(\frac{\text{kJ}}{\text{kg}}\right)$$

Specific consumption

$$= \text{mass of working substance to produce unit work} \left(\frac{\text{kg}}{\text{kJ}}\right)$$

Thermal efficiency (η_{th})

$$= \frac{\text{specific work transfer}}{\text{specific heat transfer to cycle}}$$

4.2 Reversible processes

The five processes tabulated are found useful.

Reversible non-flow processes for vapours

Name	*Law*	*Displacement work* $\int_1^2 p dv$	Δu	$q = w + \Delta u$
Constant volume process	$v = c$	0	$u_2 - u_1$	$q = (u_2 - u_1)$
Constant pressure process	$p = c$	$w = p(v_2 - v_1)$	$u_2 - u_1$	$q = p(v_2 - v_1) + (u_2 - u_1)$ or $q = h_2 - h_1$
Hyperbolic process	$pv = c$	$w = p_1 v_1 \ln\left(\frac{v_2}{v_1}\right)$	$u_2 - u_1$	$q = p_1 v_1 \ln\left(\frac{v_2}{v_1}\right) + (u_2 - u_1)$
Polytropic process	$pv^n = c$	$w = \frac{p_1 v_1 - p_2 v_2}{n - 1}$	$u_2 - u_1$	$q = \frac{p_1 v_1 - p_2 v_2}{n - 1} + (u_2 - u_1)$ $q = \frac{(h_1 - h_2) + n(u_2 - u_1)}{n - 1}$
Adiabatic process	$pv^k = c$	$w = \frac{p_1 v_1 - p_2 v_2}{k - 1}$	$u_2 - u_1$	$q = 0$

In this table the values of the relevant properties are obtained from the data programs.

4.3 Reversible non-flow processes for gases

By use of the ideal gas law ($pv = RT$) and the specific heat relations for u and h ($\Delta u = C_v \Delta T$, $\Delta h = C_p \Delta T$), and the ratio $C_p/C_v = \gamma$ for the index k in adiabatic processes, the table becomes as shown below. It

Name	*Law*	*Displacement work* $\int_1^2 p dv$	Δu	$q = w + \Delta u$
Constant volume process	$v = c$	0	$C_v(T_2 - T_1)$	$q = C_v(T_2 - T_1)$
Constant pressure process	$p = c$	$w = R(T_2 - T_1)$	$C_v(T_2 - T_1)$	$q = C_p(T_2 - T_1)$
Constant temp. (isothermal) process	$pv = c$ $T = c$	$w = RT_1 \ln\left(\frac{v_2}{v_1}\right)$	$C_v(T_2 - T_1)$	$q = RT_1 \ln\left(\frac{v_2}{v_1}\right) + C_v(T_2 - T_1)$
Polytropic process	$pv^n = c$	$w = \frac{R(T_1 - T_2)}{n - 1}$	$C_v(T_2 - T_1)$	$q = \left(\frac{\gamma - n}{\gamma - 1}\right)\frac{R(T_1 - T_2)}{(n - 1)}$
Adiabatic process	$pv^\gamma = c$	$w = \frac{R(T_1 - T_2)}{\gamma - 1}$	$C_v(T_2 - T_1)$	0

should also be noted that the hyperbolic process ($pv = c$) becomes isothermal ($T = c$).

The polytropic and adiabatic relations for a perfect gas may also be written in terms of temperature and volume

$$\frac{T_2}{T_1} = \left(\frac{p_2}{p_1}\right)^{\frac{n-1}{n}} \text{ and } T_1 v_1{}^{n-1} = T_2 v_2{}^{n-1}$$

$$\frac{T_2}{T_1} = \left(\frac{p_2}{p_1}\right)^{\frac{\gamma-1}{\gamma}} \text{ and } T_1 v_1{}^{\gamma-1} = T_2 v_2{}^{\gamma-1}$$

4.4 Steady flow processes for vapours and gases

Reversible steady flow processes can also be represented by the mathematical relations above and it is possible, provided changes in kinetic and potential energy are neglected, to evaluate shaft work transfer from

$$w_x = -\int_1^2 v\mathrm{d}p$$

This is not found to be a very useful method of analysis as the majority of practical flow processes are usually workless or ideally adiabatic, and the steady flow energy equation will yield shaft work or heat transfer directly. Four cases are of interest.

Device	$q - w_x = \Delta(h + \frac{V^2}{2} + gz)$
Heat exchanger	$q = \Delta h$ if $\Delta(\frac{V^2}{2} + gz)$ negligible
Adiabatic turbine or rotary compressor	$-w_x = \Delta h$ if $\Delta(\frac{V^2}{2} + gz)$ negligible
Adiabatic nozzle or diffuser	$0 = \Delta(h + \frac{V^2}{2})$ if Δgz negligible
Adiabatic throttle valve	$0 = \Delta h$ if $\Delta(\frac{V^2}{2} + gz)$ negligible

Since no reversible process laws have been integrated to produce these results they apply equally to reversible or irreversible flow processes. The relation between the output for reversible and irreversible cases is called the process efficiency and for adiabatic cases is discussed in section 2.8 (isentropic efficiency).

4.5 Combustion processes

As an alternative to the simple calorific value technique discussed in section 3.9, property changes in combustion processes can be determined by application of the laws of thermodynamics.

Initially the chemical equation for the reaction must be established by atom balance and a mass balance can be obtained using relative molecular mass values[1], for example

$$C_2H_4 + 3O_2 = 2\,CO_2 + 2H_2O \text{ (molal)}$$
$$28 \text{ kg} + 96 \text{ kg} = 88 \text{ kg} + 36 \text{ kg (mass)}$$

The molal equation also represents volumes (Avogadro's Law).

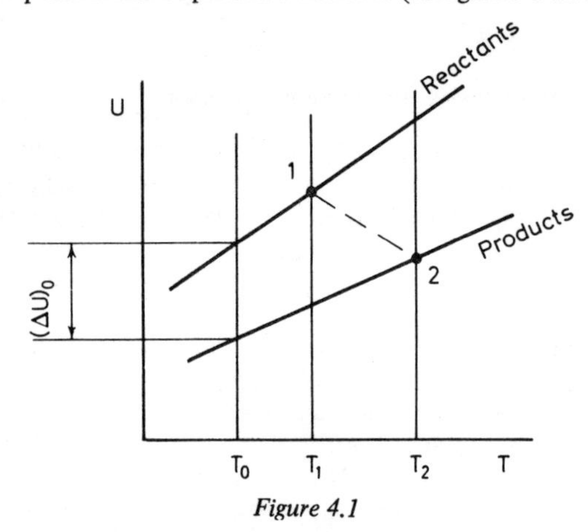

Figure 4.1

When this equation has been established the first law of thermodynamics may be applied to the reactants (R) and products (P) in a reaction in which the temperature changes from T_1 to T_2. Figure 4.1 shows a *non-flow process* in which the energy change $(\Delta U)_0$ is known. The energy change in the complete reaction is then

$$\Delta U = (U_2 - U_0)_P + (\Delta U)_0 + (U_0 - U_1)_R \qquad (4.1)$$

Similarly in a *flow process* (assuming all the reactants at the same temperature)

$$\Delta H = (H_2 - H_0)_P + (\Delta H)_0 + (H_0 - H_1)_R \qquad (4.2)$$

Values of $(\Delta U)_0$ and $(\Delta H)_0$ are determined from basic chemical data for the *enthalpy of formation* (Δh_f) of substances. The enthalpy of formation of a substance is the enthalpy of the product of a constant pressure, isothermal reaction in which the substance is formed from its elements.

Calorific value (section 3.9) avoids this complex process.

4.6 Combustion of hydrocarbon fuels

In most applications the chemical equation concerned is the combustion of a hydrocarbon fuel in air. The engineer is interested in the *stoichiometric* air–fuel ratio so that the fuel just burns with no excess oxygen, and also in combustion processes having air-fuel ratios with a small excess of oxygen to ensure complete burning of the fuel. It is possible to analyse the combustion products and then to determine the fuel analysis and the air–fuel ratio used. The necessary chemical equation is written

$$aC + bH_2 + 21xO_2 + 79\,xN_2 = dCO_2 + eO_2 + gN_2 + kH_2O$$

where a, b, d, e, g, and k are volumetric or molal percentages; i.e. $b = (100 - a)$, $(d + e + g) = 100$, since k is not usually determined in exhaust analysis. It has been assumed that adequate air of volumetric proportion 21 per cent oxygen, 79 per cent nitrogen is supplied (there is excess oxygen) to completely burn all the fuel. In such problems the values of d, e and g will be known and four chemical balance equations (C, H_2, O_2, N_2) are used to solve for a, b, X and k. The quantity 100X is the number of moles of air supplied to burn a moles of carbon and b moles of oxygen. The air–fuel ratio by mass is then given by

$$\text{a f r} = \frac{2900X}{12a + 2b}$$

The stoichiometric air–fuel ratio is obtained by solving the equation with $e = 0$, a and b only being known, and the excess air is defined by

$$\text{Excess air} = \frac{\text{afr} - \text{afr (stoic)}}{\text{afr(stoic)}} \times 100\%$$

This method will also suffice if there is a deficiency of air by substituting eCO for eO_2 on the right-hand side of the equation, as it is normally assumed that the hydrogen burns completely leaving the carbon with the oxygen deficiency.

4.7 Ideal cycles

The ideal cycles below are considered equivalent to those used in real machines. The use of process efficiency to allow for real process irreversibility is indicated. The ideal reversible heat engine discussed in section 2.7 would use a cycle with constant temperature heat transfer processes to achieve its ideal (Carnot) efficiency. It is known as the Carnot cycle. It is not found practical to use constant temperature heat transfer processes (unless a simple phase change is the sole occurrence) and ideal practical heat transfer processes are made at constant pressure. This means that no practical cycle using such processes can expect to achieve a thermal efficiency equal to the Carnot efficiency. (Similar arguments apply to reversed heat engine cycles.)

4.8 Steam plant cycle

The ideal steam plant cycle is called the *Rankine cycle.* Figure 4.2 shows the superheat cycle and the associated plant. It is a constant pressure flow process cycle.

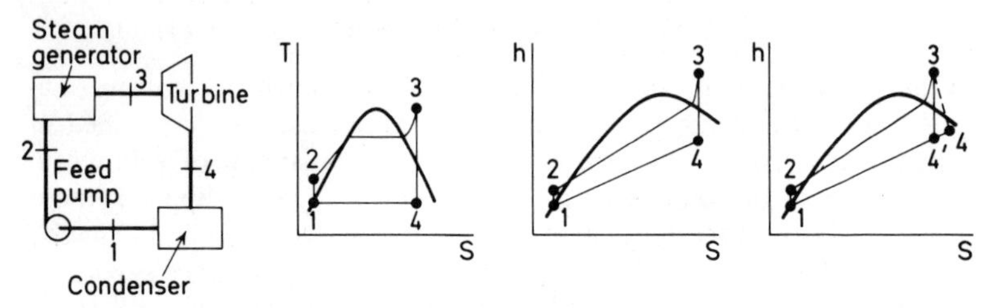

Figure 4.2 Steam plant cycle

Process	*Description*	$q - w_x = \Delta h$, $\Delta(\frac{V^2}{2} + gz)$ *negligible*
1–2	Reversible, adiabatic compression of saturated water in a *feed pump*	$-w_x = \Delta h = h_2 - h_1$
2–3	Reversible, constant pressure heat transfer to cycle in a *steam generator*	$q = \Delta h = h_3 - h_2$
3–4	Reversible, adiabatic expansion of steam in a *turbine*	$-w_x = \Delta h = h_4 - h_3$ or $w_x = h_3 - h_4$
4–1	Reversible, constant pressure heat transfer from cycle in a *condenser*	$q = \Delta h = h_1 - h_4$

The feed pump work $(h_2 - h_1)$ is considered negligible and thus h_2 is considered equal to h_1, hence

$$\text{specific work} = h_3 - h_4 \tag{4.3}$$

$$\text{thermal efficiency} = \frac{h_3 - h_4}{h_3 - h_1} \tag{4.4}$$

The effect of irreversibility is to reduce the work transfer in the turbine and the isentropic efficiency is given by (Figure 4.2)

$$\eta_{\text{isen}} = \frac{\text{actual work}}{\text{ideal work}} = \frac{h_3 - h_4}{h_3 - h_4'}$$

Condenser pressure is dictated by the natural coolant available at perhaps 19°C, and the steam will condense at about 29°C at a pressure of 0.004 MN/m^2.

4.9 Gas turbine plant cycle

The ideal gas turbine plant cycle is called the *Joule cycle.* Figure 4.3 shows the cycle and associated plant. It is a constant pressure flow process cycle

Process	*Description*	$q - w_x = \Delta h$, $\Delta(\frac{V^2}{2} + gz)$ *negligible*
1–2	Reversible, adiabatic compression in a *rotary compressor*	$-w_x = h_2 - h_1 = C_p (T_2 - T_1)$
2–3	Reversible, constant pressure heat transfer to cycle in a *heater*	$q = h_3 - h_2 = C_p (T_3 - T_2)$
3–4	Reversible, adiabatic expansion in a *turbine*	$-w_x = h_4 - h_3$ $w_x = C_p (T_3 - T_4)$
4–1	Reversible, constant pressure heat transfer from cycle in a *cooler*	$q = h_1 - h_4 = C_p (T_1 - T_4)$

$$\text{specific work} = C_p (T_3 - T_4) - C_p(T_2 - T_1) \tag{4.5}$$

$$\text{thermal efficiency} = \frac{C_p(T_3 - T_4) - C_p(T_2 - T_1)}{C_p(T_3 - T_2)} \tag{4.6}$$

The effect of irreversibility is to reduce the work transfer in the turbine and to increase the work transfer in the compressor. The isentropic efficiencies are given by (Figure 4.3)

$$\eta_t = \frac{\text{actual work}}{\text{ideal work}} = \frac{T_3 - T_4}{T_3 - T_4'}$$

and

$$\eta_c = \frac{\text{ideal work}}{\text{actual work}} = \frac{T_2' - T_1}{T_2 - T_1}$$

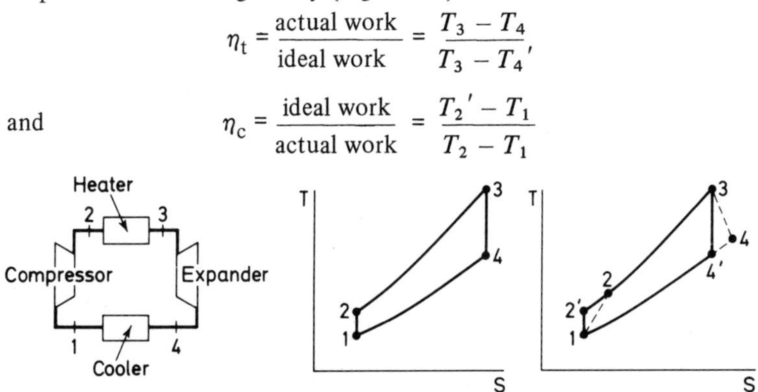

Figure 4.3 Joule cycle

4.10 Reciprocating engine cycles

The ideal cycle for the spark ignition and compression ignition engine is called the *Otto cycle.* Figure 4.4 shows the cycle and two modifications, known as the Diesel and dual cycles, to allow for differing combustion processes in real engines. They are non-flow process cycles in

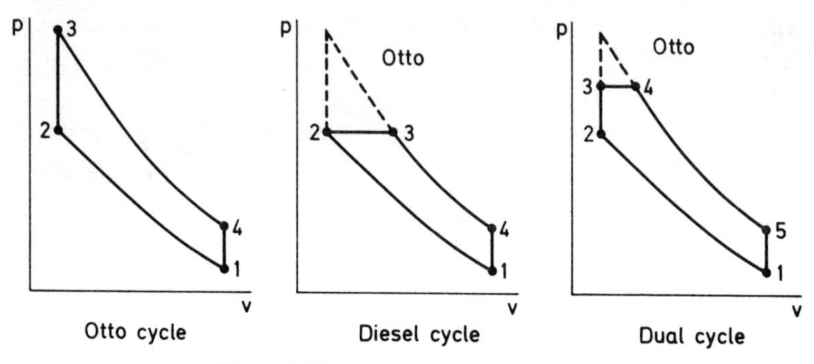

Figure 4.4 Reciprocating engine cycles

piston and cylinder devices with constant volume, or constant pressure heat transfer processes.

Process	*Description (Otto cycle)*	$q - w = \Delta u$
1–2	Reversible, adiabatic compression	$-w = (u_2 - u_1)$ $= C_v(T_2 - T_1)$
2–3	Reversible, constant volume heat transfer to cycle	$q = (u_3 - u_2)$ $= C_v(T_3 - T_2)$
3–4	Reversible, adiabatic expansion	$-w = (u_4 - u_3)$ $w = C_v(T_3 - T_4)$
4–1	Reversible, constant volume heat transfer from cycle	$q = (u_1 - u_4)$ $= C_v(T_1 - T_4)$

$$\text{specific work} = C_v(T_3 - T_4) - C_v(T_2 - T_1) \tag{4.7}$$

$$\text{thermal efficiency} = \frac{C_v(T_3 - T_4) - C_v(T_2 - T_1)}{C_v(T_3 - T_2)} \tag{4.8}$$

For the Diesel cycle these expressions become

$$\text{specific work} = R(T_3 - T_2) + C_v(T_3 - T_4) - C_v(T_2 - T_1) \tag{4.9}$$

$$\text{thermal efficiency} = \frac{R(T_3 - T_2) + C_v(T_3 - T_4) - C_v(T_2 - T_1)}{C_p(T_3 - T_2)} \tag{4.10}$$

For the dual cycle these expressions become

$$\text{specific work} = R(T_4 - T_3) + C_v(T_4 - T_5) - C_v(T_2 - T_1) \quad (4.11)$$

$$\text{thermal efficiency} = \frac{R(T_4 - T_3) + C_v(T_4 - T_5) - C_v(T_2 - T_1)}{C_p(T_4 - T_3) + C_v(T_3 - T_2)} \quad (4.12)$$

The Otto cycle has the highest efficiency of the three ideal cycles if they use the same compression ratio (V_2/V_1), but in practice, due to the combustion methods used, the dual cycle in the high speed compression ignition engine is able to use a higher compression ratio and so achieve higher efficiency.

4.11 Reciprocating air compression cycle

The ideal cycle for a reciprocating air compressor is shown in Figure 4.5. It is a non-flow process cycle.

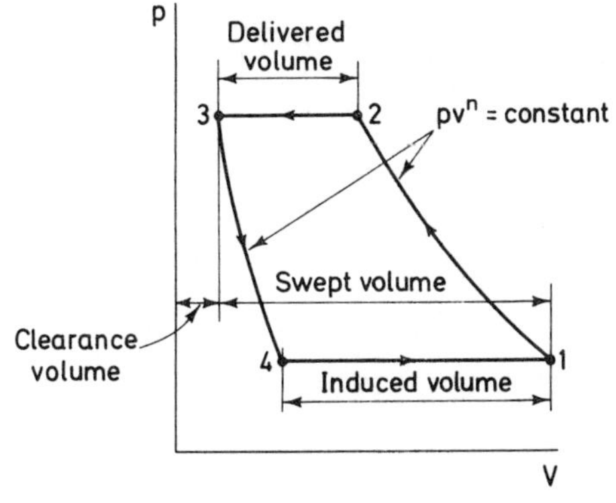

Figure 4.5 Reciprocating compressor cycle

Process	*Description*	*Displacement work transfer* = $\int_1^2 p\,dV$
1–2	Reversible, polytropic compression, $pV^n = C$	$W = \dfrac{(p_1 V_1 - p_2 V_2)}{n-1}$
2–3	Delivery of compressed air at constant pressure and temperature	$W = p_2(V_3 - V_2)$
3–4	Reversible, polytropic expansion, $pV^n = C$	$W = \dfrac{(p_3 V_3 - p_4 V_4)}{n-1}$
4–1	Induction of fresh air at constant pressure and temperature	$W = p_1(V_1 - V_4)$

By expressing these work transfers in terms of m the delivered (and induced) mass and m_c the clearance mass it is found that m_c disappears and the total work per cycle becomes, provided that the index n is the same for compression and expansion,

$$W = m\left\{\frac{n}{n-1}\right\}R(T_1 - T_2) \text{ or } W = m\left\{\frac{n}{n-1}\right\}RT_1\left(1-\left\{\frac{p_2}{p_1}\right\}^{\frac{n-1}{n}}\right) \quad (4.13)$$

The induced mass m is more usually quoted as a volume flow rate measured at some specified state.

4.12 Vapour compression refrigeration and heat pump cycle

The simple vapour compression *refrigeration* cycle is shown in Figure 4.6. It is a constant pressure, flow process cycle.

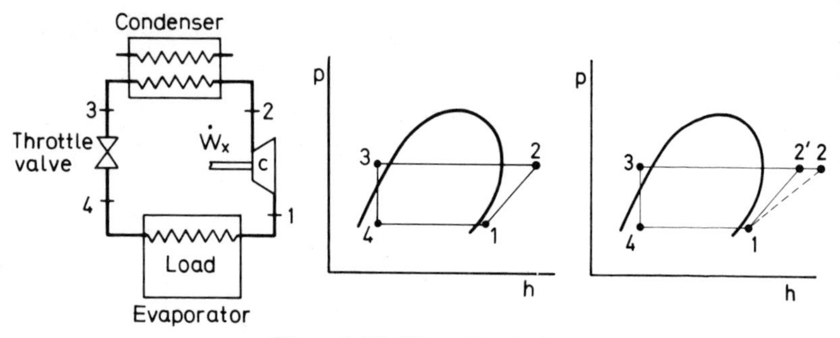

Figure 4.6 Refrigeration cycle

Process	*Description*	$q - w_x = \Delta h$, $\Delta(\frac{V^2}{2} + gz)$ *negligible*
1–2	Reversible, adiabatic compression of a superheated vapour in a *compressor*	$-w_x = h_2 - h_1$
2–3	Reversible constant pressure heat transfer from the cycle in a *condenser*	$q = h_3 - h_2$
3–4	*Irreversible* expansion of a vapour in a *throttle valve*	$h_3 = h_4$
4–1	Reversible constant pressure heat transfer to the cycle in an *evaporator*	$q = h_1 - h_4$

$$\text{refrigeration effect} = h_1 - h_4 \quad (4.14)$$

$$\text{coefficient of performance} = \frac{h_1 - h_4}{h_2 - h_1} \quad (4.15)$$

The effect of irreversibility is to increase the work transfer needed in the compression process (Figure 4.6) and the isentropic efficiency is given by

$$\eta_{isen} = \frac{\text{ideal work}}{\text{actual work}} = \frac{h_2' - h_1}{h_2 - h_1}$$

Evaporator pressure is dictated by the temperature to which the refrigerator contents are to be cooled. Condenser pressure is determined by the natural coolant temperature of 19°C so that condensing will take place at approximately 30°C.

The *heat pump* cycle is identical except that the output is the condenser heat transfer and this is called the *specific heating effect.*

$$\text{specific heating effect} = h_2 - h_3 \tag{4.16}$$

$$\text{coefficient of performance} = \frac{h_2 - h_3}{h_2 - h_1} \tag{4.17}$$

Evaporator conditions are dictated by the temperature of the energy source used and condenser conditions will be chosen to suit the temperature required in the space being heated.

4.13 The air conditioning 'cycle'

The processes forming the 'cycle' for air conditioning plant are shown in Figure 4.7. The cycle is a series of flow processes.

Process	*Description*	$q - w_x = \Delta h$, $\Delta(\frac{V^2}{2} + gz)$ *negligible*
1–2–3	Mixing of m_1 of recirculated air at state 1 with m_2 of fresh air at state 2	$h_3 = \dfrac{m_1 h_1 + m_2 h_2}{m_1 + m_2}$
3–4–5	Cooling of air at constant pressure until at the dew point (4) condensation occurs. Continued cooling causes vapour to condense ($\dot{m}_{cond}$) and reduce the humidity where	$q_{35} = (h_5 - h_3) + \left(\dfrac{\dot{m}_{cond}}{\dot{m}_{dry\ air}}\right) h_{f_5}$ $\left(\dfrac{\dot{m}_{cond}}{\dot{m}_{dry\ air}}\right) = \omega_3 - \omega_5$
5–6	Reheating at constant pressure of the cold dehumidified air to the delivery state 6	$q_{56} = h_6 - h_5$
6–1	The load causes the conditioned air at state 6 to become the maintained state air at state 1	

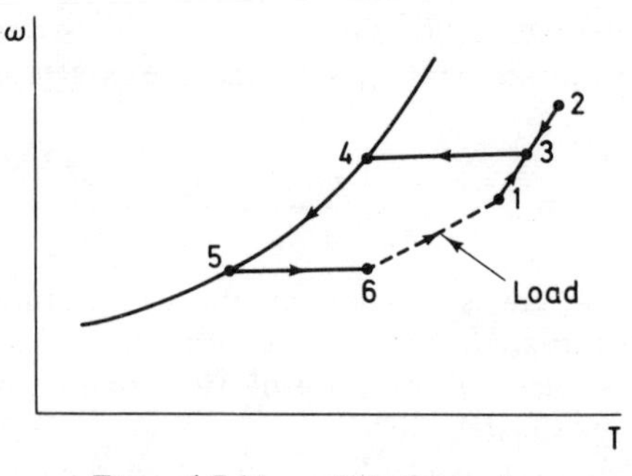

Figure 4.7 Air conditioning cycle

4.14 Reference

1. Haywood, R.W., *Thermodynamic Tables in S.I. (Metric) Units*, Cambridge University Press (1968).

WORKED EXAMPLES

Example 4.1 Isentropic process

Write a program to determine a value for n in a relation of the form pv^n = constant describing a reversible constant entropy process for superheated steam from 5 MN/m^2, 600°C to 0.5 MN/m^2. Include in the program a provision to determine Δh in the process.

The necessary reading for this example is in section 4.2.

```
LIST
 10  PRINT "EXAMPLE 4.1"
 20  PRINT
 30  DIM P[120],T[120],H[120],S[120],U[120],V[120]
 40  FOR N=0 TO 119
 50   READ P[N],T[N],H[N],S[N],U[N],V[N]
 60  NEXT N
 70  N=119
 80  S1=S[N]
 90  H1=H[N]
 100  V1=V[N]
 110  B=0
 120  C=0
 130  FOR J=80 TO 89
 140   X=1
 150   FOR N=80 TO 89
 160    IF N=J THEN GOTO 180
 170    X=X*(S1-S[N])/(S[J]-S[N])
 180   NEXT N
 190   B=B+X*V[J]
 200   C=C+X*H[J]
 210  NEXT J
 220  V2=B
 230  H2=C
 240  RP=5/0.5
 250  RV=V2/V1
 260  D=LOG[RP]
```

```
270  E=LOG[RV]
280  K=D/E
290  PRINT "ADIABATIC INDEX FOF SUPERHEATED STEAM IS";K
300  DH=H2-H1
310  PRINT "ISENTROPIC ENTHALPY CHANGE IS";DH;"KJ/KG"
320  PRINT "DH IS ALSO INTEGRAL VDP"
330  VDP=K*(500*V2-5000*V1)/(K-1)
340  PRINT "ISENTROPIC ENTHALPY CHANGE BY INTEGRAL VDP";VDP;"KJ/KG"
350  END


RUN
EXAMPLE 4.1

ADIABATIC INDEX FOF SUPERHEATED STEAM IS 1.28595156
ISENTROPIC ENTHALPY CHANGE IS -711.328414KJ/KG
DH IS ALSO INTEGRAL VDP
ISENTROPIC ENTHALPY CHANGE BY INTEGRAL VDP -708.205022KJ/KG

STOP AT 350
```

Program nomenclature

RP pressure ratio
RV volume ratio
DH specific enthalpy decrease
VDP $\int v\, dp$

Program notes

(1) Steam table data is read in at lines 40 to 60.
(2) Line 70 selects the appropriate line in the data (by setting N = 119) for the initial state values.
(3) Lines 110 to 210 interpolate in the appropriate pressure block to determine final state values.
(4) Lines 220 to 280 find K from $p_1 v_1{}^K = p_2 v_2{}^K$ by taking logarithms.
(5) Line 300 finds Δh by using

$$T\mathrm{d}s = \mathrm{d}h - v\mathrm{d}p$$

but for an isentropic process $\mathrm{d}s = 0$, so that

$$\mathrm{d}h = v\mathrm{d}p$$

For the process $pv^K = c$

$$\Delta h = \int_1^2 v\mathrm{d}p = \frac{K}{K-1}\,(p_2 v_2 - p_1 v_1)$$

Example 4.2 Combustion process

Write a program to find the heat transfer rate when 1 kmol/s of methane burns at constant pressure in a steady flow of oxygen to form H_2O and CO_2 with no excess oxygen. The initial temperature of the reactants is T_1 and the final temperature of the products is T_2. Run the program with T_1 = 25°C and T_2 = 25°C; T_1 = 20°C and T_2 = 25°C and 250°C. (Assume H_2O is in the vapour phase.)

Data:

	Specific enthalpy of formation (Δh_f) at 298K, kJ/kmol	*Specific heat capacity (C_p) kJ/kg K*
Methane	−74 870	2.23
Steam	−241 830	1.865
Carbon dioxide	−393 520	0.843
Oxygen	0	0.918

The necessary reading for this example is in section 4.5.

```
LIST
 10  PRINT "EXAMPLE 4.2"
 20  PRINT "ENTHALPY CHANGE IN REACTION AT 298 K=SUM(N*HF298)"
 30  REM   LET HM BE METHANE HF298
 40  REM    LET HO BE OXYGEN HF298
 50  REM    LET HCD BE CARBON DIOXIDE HF298
 60  REM    LET HW BE STEAM HF298
 70  HM=-74870
 80  HW=-241830
 90  HCD=-393520
 100  HO=0
 105  REM  HF298 IN KJ/KG MOL
 110  HR=(HM+HO)
 120  HP=(HCD+2*HW)
 130  H2=(HP-HR)
 140  PRINT "ENTHALPY CHANGE OF REACTANTS OR PRODUCTS=SUM(N*CP*DT)"
 150  REM    LET CPM BE SPECIFIC HEAT OF METHANE
 160  REM    LET CPO BE SPECIFIC HEAT OF OXYGEN
 170  REM    LET CPW BE SPECIFIC HEAT OF STEAM
 180  REM    LET CCD BE SPECIFIC HEAT OF CARBON DIOXIDE
 185  REM SPECIFIC HEAT IN KJ/KG K,NEED TO MULTIPLY BY MOLECULAR MASS
 190  CPM=2.23
 200  CPO=0.918
 210  CPW=1.865
 220  CCD=0.843
 230  PRINT "REACTANTS"
 240  PRINT "TEMPERATURE?"
 250  INPUT T1
 260  H1=((1*CPM*16)+(2*CPO*32))*(298-(T1+273))
 270  PRINT "PRODUCTS"
 280  PRINT "TEMPERATURE?"
 290  INPUT T2
 300  H3=((1*CCD*44)+(2*CPW*18))*((T2+273)-298)
 310  H=(H1+H2+H3)
 320  PRINT "HOW MANY MOLS  PER SECOND OF METHANE?"
 330  INPUT N
 340  Q=N*H
 350  PRINT "HEAT TRANSFER RATE=";Q;"KW"
 360  END
```

```
RUN
EXAMPLE 4.2
ENTHALPY CHANGE IN REACTION AT 298 K=SUM(N*HF298)
ENTHALPY CHANGE OF REACTANTS OR PRODUCTS=SUM(N*CP*DT)
REACTANTS
TEMPERATURE?
? 25
PRODUCTS
TEMPERATURE?
? 25
HOW MANY MOLS  PER SECOND OF METHANE?
? 1
HEAT TRANSFER RATE= -802310KW

STOP AT 360


RUN
EXAMPLE 4.2
ENTHALPY CHANGE IN REACTION AT 298 K=SUM(N*HF298)
ENTHALPY CHANGE OF REACTANTS OR PRODUCTS=SUM(N*CP*DT)
REACTANTS
TEMPERATURE?
? 20
PRODUCTS
TEMPERATURE?
? 25
HOW MANY MOLS  PER SECOND OF METHANE?
? 1
HEAT TRANSFER RATE= -801837.840KW

STOP AT 360


RUN
EXAMPLE 4.2
ENTHALPY CHANGE IN REACTION AT 298 K=SUM(N*HF298)
ENTHALPY CHANGE OF REACTANTS OR PRODUCTS=SUM(N*CP*DT)
REACTANTS
TEMPERATURE?
? 20
PRODUCTS
TEMPERATURE?
? 250
HOW MANY MOLS  PER SECOND OF METHANE?
? 1
HEAT TRANSFER RATE= -778385.640KW

STOP AT 360
```

Program nomenclature

HF enthalpy of formation
Q heat transfer rate

Program notes

(1) The reaction is

$$CH_4 + 2O_2 = CO_2 + 2H_2O$$

The enthalpy change in the isothermal reaction is found from enthalpy of formation data in lines 70 to 130.

(2) The enthalpy change of the reactants and products is found in lines 140 to 300.

(3) The heat transfer rate in the non-isothermal reaction is found in lines 310 to 350 by summing the enthalpy change of the reactants, the enthalpy change in the isothermal reaction and the enthalpy change of the products, and using the steady flow equation

$$\dot{Q} = (\Sigma_{out} - \Sigma_{in})\dot{m}h$$

(4) This program is not suitable for use at higher temperatures unless the values of specific heat are changed. It should never be used in adiabatic or other high temperature reactions which involve chemical dissociation.

Example 4.3 Joule cycle

In the Joule cycle with allowance for compressor and turbine isentropic efficiencies η_c and η_t the specific work transfer can be shown to be

$$w = \eta_t C_p T_3 \left(1 - \frac{1}{x}\right) - \frac{C_p T_1}{\eta_c}(x - 1)$$

where $x = r_p^{\frac{\gamma-1}{\gamma}}$, r_p being the cycle pressure ratio and T_3 and T_1 the cycle maximum and minimum temperatures. Write a program to evaluate specific work transfer with r_p = 2, 4, 6, 8, 10 and 12 and T_1 = 300 K, T_3 = 1500 K, η_c = 0.8, η_t = 0.85, C_p = 1.01 kJ/kg K, γ = 1.4. Plot the results and compare the pressure ratio for maximum specific work with that obtained by differentiation of the equation above for w.

The necessary reading for this example is in section 4.9.

```
LIST
10  PRINT "EXAMPLE 4.3"
20  DIM RP[6],W[6]
30  FOR N=0 TO 5
40   READ RP[N]
50  NEXT N
60  FOR N=0 TO 5
70   G=1.4
80   I=(G-1)/G
90   RP=RP[N]
100   X=RP↑I
110   ET=0.85
120   EC=0.8
130   CP=1.01
140   T3=1500
150   T1=300
160   W[N]=(ET*CP*T3*(1-(1/X)))-(CP*T1)*(X-1)/EC
170  NEXT N
180  PRINT "PRESSURE RATIO SPECIFIC WORK(KJ/KG)"
190  PRINT
200  FOR N=0 TO 5
```

```
210   PRINT RP[N],W[N]
220  NEXT N
230  DATA 2,4,6,8,10,12
240  REM BY CHANGING THE DATA AFTER A RUN THE MAXIMUM CAN BE MORE CLOSELY IDENTIFIED
250  END

RUN
EXAMPLE 4.3
PRESSURE RATIO SPECIFIC WORK(KJ/KG)

 2               148.411724
 4               237.088176
 6               262.755481
 8               269.518689
 10              268.261383
 12              263.014656

STOP AT 250
```

Program nomenclature

RP	pressure ratio
W	work transfer
G	gamma
ET	turbine isentropic efficiency
EC	compressor isentropic efficiency
CP	specific heat
T3	maximum cycle temperature
T1	minimum cycle temperature

Program notes

(1) This program is structured differently. It requires no interactive communication as it solves one equation.
(2) It could be modified to vary the constants by arranging a series of inputs for η_t, η_c, T_1 and T_3 so that a family of curves could be produced. These changes should be arranged so that only one variable is altered at any one time in order to observe the significance of this particular parameter. Try changing T_3 to 1600 K, then change η_t to 0.9 with T_3 at 1500 K, then 1600 K, etc.
(3) Show that the differentiation results in the pressure ratio for maximum work output being given by

$$r_p = \left[\frac{\eta_c \eta_t T_3}{T_1}\right]^{\frac{\gamma}{2(\gamma-1)}}$$

and evaluate this for the given data.
(4) A plot routine could be employed if available for the single curve or for families of curves when one of the four variables is changed, the other three being held constant.

Example 4.4 Refrigeration cycle

Write a program to determine the coefficient of performance of a refrigeration cycle using refrigerant 12. Run the program for a refrigerator with evaporation temperature –5°C and condenser temperature 30°C, for a freezer with the evaporator temperature lowered to –50°C and for a water chiller with the evaporator temperature raised to 5°C. Assume that the refrigeration cycle has reversible compression with the vapour dry saturated at compressor entry and with the liquid at condenser exit in the saturated state.

The necessary reading for this example is in section 4.12.

```
LIST
 10  PRINT "EXAMPLE 4.4"
 20  DIM T[100],P[100],V[100],H[100],S[100]
 30  FOR N=0 TO 99
 40   READ T[N],P[N],V[N],H[N],S[N]
 50  NEXT N
 60  PRINT "FIND H4"
 70  N=16
 80  H4=H[N]
 90  M=0
 100  FOR M=0 TO 2
 110   PRINT "FIND H1,S1"
 120   PRINT "WHAT IS N?"
 130   REM INPUT D IS N PUT D= 21 THEN 29 THEN 31
 140   INPUT D
 150   N=D
 160   H1=H[N]
 170   S1=S[N]
 180   B=0
 190   PRINT "WHAT IS J RANGE TO FIND H2?"
 200   PRINT "NO INPUT REQUIRED,SEE LINE 205"
 205   REM INTERPOLATION IN LINES 88 TO 90 IN ALL CASES
 210   S=S1
 220   FOR J=88 TO 90
 230    X=1
 240    FOR N=88 TO 90
 250     IF N=J THEN GOTO 270
 260     X=X*(S-S[N])/(S[J]-S[N])
 270    NEXT N
 280    B=B+X*H[J]
 290   NEXT J
 300   H2=B
 310   COP=(H1-H4)/(H2-H1)
 320   PRINT "COEFFICIENT OF PERFORMANCE";COP
 330   PRINT
 340  NEXT M
 350 END
```

```
RUN
EXAMPLE 4.4
FIND H4
FIND H1,S1
WHAT IS N?
? 21
WHAT IS J RANGE TO FIND H2?
NO INPUT REQUIRED,SEE LINE 205
COEFFICIENT OF PERFORMANCE 1.93430472
```

```
NO INPUT REQUIRED,SEE LINE 205
COEFFICIENT OF PERFORMANCE 6.55709546

FIND H1,S1
WHAT IS N?
? 31
WHAT IS J RANGE TO FIND H2?
NO INPUT REQUIRED,SEE LINE 205
COEFFICIENT OF PERFORMANCE 9.92549827

STOP AT 350
```

Program nomenclature

COP coefficient of performance

Program notes

(1) To determine the coefficient of performance three enthalpy values are required: in this case one *g* value, one *f* value and one superheat value. Three 'interpolations' are needed (Figure 4.6).
(2) Data is read in lines 30 to 50. All inputs in this case are exact lines in the tables which enables interpolation to be reduced by choice of N.
(3) The three cases are within the M loop of lines 90 to 340 except for h_4 which remains constant. The loop requires inputs to set N and the input values are in the REM statements.
(4) The results show that as the cycle is required to perform over a larger temperature range, that is to freeze rather than refrigerate and to refrigerate rather than chill, so the coefficient of performance falls. This means that more power input is required to freeze than to refrigerate and more to refrigerate than to chill.

PROBLEMS

(4.1) Write a program to determine a value for n in a relation of the form pv^n = constant describing a reversible constant entropy process for steam that is initially dry saturated. Choose suitable initial and final states.
(4.2) In an Orsat test on the combustion products of a hydrocarbon fuel burned in air the volumetric analysis of carbon dioxide, oxygen and carbon monoxide is determined. The balance of the products is assumed to be nitrogen and air is made up of 21 per cent oxygen and 79 per cent nitrogen by volume.

Write a programe to find the air–fuel ratio used in the combustion and the mass analysis of the fuel. Run the program with the following test results.

CO_2 (%)	7.5	7.2	6.5
O_2 (%)	9.4	0	0
CO (%)	0	0	1

(4.3) Modify the program of example 4.3 to determine the thermal efficiency of the cycle, then run the program to determine the pressure ratio for maximum thermal efficiency by hand plotting (as in example 4.3) and compare the value of the pressure ratio for maximum thermal efficiency with that for maximum work obtained in example 4.3. First show that

$$\eta_{th} = \frac{w}{C_p\left[T_3 - \left\{\frac{T_1(x-1)}{\eta_c} + T_1\right\}\right]}$$

(4.4) (a) A Rankine cycle using superheated steam has a condenser pressure of 0.004 MN/m^2 and a steam generator pressure of 1 MN/m^2. Write a program to determine the thermal efficiency (ignoring feed pump work) with maximum cycle temperatures of 225, 300, 400, 500 and 600°C. Plot the results.
(b) Holding the maximum cycle temperature constant at 600°C, modify the program to vary the steam generator pressure. Evaluate this program at 1, 2, 3, 4 and 5 MN/m^2. Plot the results. What is the dryness fraction at turbine exit at each pressure?
(4.5) Write a program to determine the mass of air induced and delivered by a reciprocating air compressor with a clearance volume of x of the swept volume, air inlet pressure p_s and a delivery pressure p_d. The air inlet temperature is T_s. Evaluate the program with the ratio p_d/p_s = 4, 8, 12, 16, 20, etc., until the ratio p_d/p_s reaches its limiting value when the mass flow rate is zero:

$$\frac{p_d}{p_s} = \left[\frac{1+x}{x}\right]^n$$

Choose $x = 0.05$ and $n = 1.2$ with $p_s = 100$ kN/m^2 and $T_s = 300$ K.
(4.6) Write a program to determine the thermal efficiency of the Otto cycle (using air as a working substance) with compression ratio $r_v = V_1/V_2$. Include an option, or write a separate program, for the diesel cycle with the same compression ratio. For the diesel cycle the cut-off ratio, $r_c = V_3/V_2$. Evaluate the program with $r_v = 18$, $r_c = 1.96$, then increase r_v to 25.

Hint: Refer to Figure 4.4 and either rearrange the efficiency expressions in terms of r_v and r_c or determine the work and heat transfers by temperatures determined from r_v and r_c.
(4.7) Using the necessary parts of the program in problem 3.5, write a program to determine the heat transfer rate in the air conditioning

'cycle' (section 4.13) cooling process 3—4—5. Evaluate the specific heat transfer for an entry state 3 of T_{db} = 30°C, T_{wb} = 27°C and an exit state 6 of T_{db} = 20°C, T_{wb} = 16°C. To avoid steam table interpolation put $h_{f_5} = 4.2\ T_5$ kJ/kg.

If problem 3.5 has not been solved, the program of example 3.7 may be used with air entry state of T_{db} = 30°C, ϕ = 80% and an exit state of T_{db} = 20°C, ϕ = 67%.

(4.8) Repeat example 4.4 using the data block prepared for ammonia in problem 3.3. Compare the performance of the ammonia plant with that of the refrigerant 12 plant and determine the ratio of the volumes of working substance at compressor inlet for the same cooling load.

Chapter 5

Fluid flow

ESSENTIAL THEORY

5.1 Stagnation or total head properties

If a perfect gas is brought to rest reversibly, adiabatically and worklessly, the pressure and temperature will rise from the static values (p and T) to the stagnation or total head values (p_t and T_t), where

$$\frac{T_t}{T} = 1 + \frac{V^2}{2C_pT} \quad \text{and} \quad \frac{p_t}{p} = \left\{\frac{T_t}{T}\right\}^{\frac{\gamma}{\gamma - 1}} \tag{5.1}$$

These relations may also be expressed in terms of the flow Mach number, Ma.

$$\text{Ma} = \frac{\text{local velocity}}{\text{local sonic velocity}} = \frac{V}{\sqrt{\gamma RT}}$$

$$\frac{T_t}{T} = 1 + \frac{(\gamma - 1)}{2}\text{Ma}^2 \quad \text{and} \quad \frac{p_t}{p} = \left\{1 + \frac{(\gamma - 1)}{2}\text{Ma}^2\right\}^{\frac{\gamma}{\gamma-1}} \tag{5.2}$$

If a thermometer is placed in a moving stream it will record the stagnation temperature. Stagnation pressure is recorded with a pitot tube. Velocity can be determined with a pitot static tube which measures the difference between stagnation pressure and static pressure (Figure 5.1).

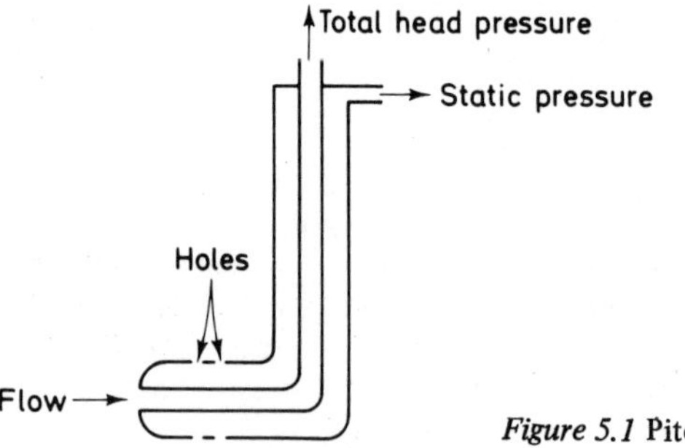

Figure 5.1 Pitot static tube

5.2 Adiabatic duct flow

In flow along a duct there will be friction which will mean that, although the stagnation temperature will be constant in the duct, there will be a fall in stagnation pressure along the duct in the direction of

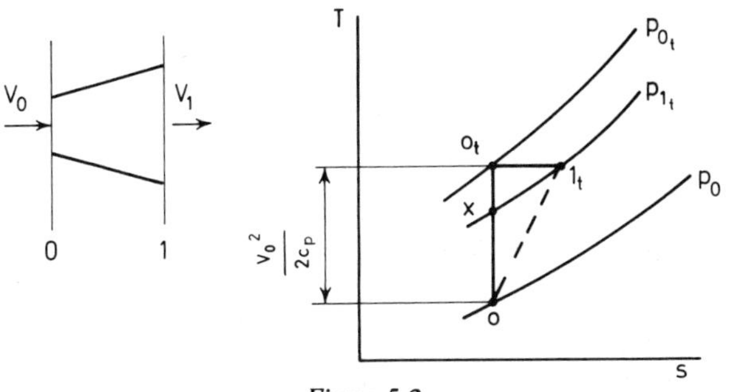

Figure 5.2

flow. For example, in a duct designed to reduce velocity and increase pressure, there will be a duct efficiency. It can be seen in Figure 5.2 that $p_{1t} < p_{0t}$ and the actual pressure achieved can be found from

$$\eta_{\text{duct}} = \frac{T_x - T_0}{T_{0t} - T_0} \quad \text{to give } T_x$$

$$\text{then} \quad \frac{p_{1t}}{p_0} = \left\{\frac{T_x}{T_0}\right\}^{\frac{\gamma}{\gamma - 1}}$$

By combining the steady flow energy equation with the continuity equation duct shape can be related to flow velocity

$$\frac{\mathrm{d}A}{A} = \frac{\mathrm{d}p}{\rho}\left\{\frac{1}{V^2} - \frac{1}{a^2}\right\}$$

where a is the local sonic velocity, A the duct cross-sectional area, V the local velocity, p and ρ pressure and density. Interpretation of this relation shows that when

$$\mathrm{d}A = 0, V = a$$

A *nozzle* which accelerates flow from subsonic to sonic flow will be convergent followed by a divergent portion to accelerate flow from sonic to supersonic. A *diffuser* which decelerates flow from supersonic to sonic will be convergent followed by a divergent portion to decelerate flow from sonic to subsonic (Figure 5.3).

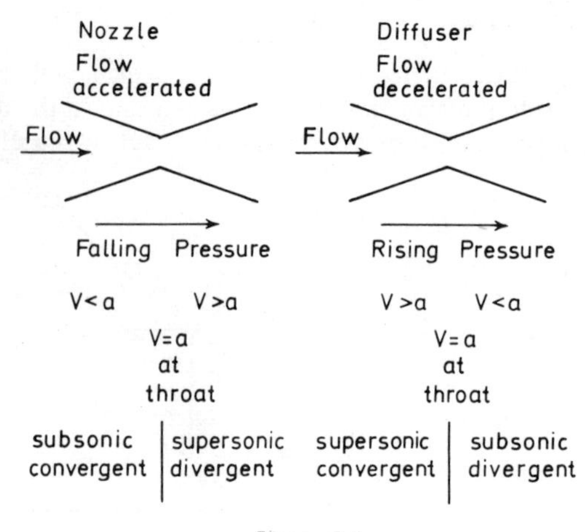

Figure 5.3

5.3 Critical pressure ratio of a nozzle

The pressure ratio in a duct $r_p = \dfrac{\text{local pressure}}{\text{inlet pressure}}$ is used to determine whether flow will become sonic at the *throat* of a convergent or convergent-divergent nozzle. It is found that a minimum value of this pressure ratio

$$r_p = \left\{\frac{2}{\gamma + 1}\right\}^{\frac{\gamma}{\gamma - 1}} \tag{5.3}$$

known as the critical pressure ratio, occurs at the *throat* of such a nozzle when the velocity is sonic ($V = a$) at the throat. There is no way of increasing the velocity beyond this local sonic value; the nozzle is said to be *choked* and is passing its maximum mass flow rate

$$V_{\text{throat}} = \sqrt{\gamma\left\{\frac{p}{\rho}\right\}_{\text{throat}}} \tag{5.4}$$

At any other cross-section 2

$$V_2 = \sqrt{\left(\frac{2\gamma}{\gamma - 1}\right)\frac{p_1}{\rho_1}\left\{1 - \left(\frac{p_2}{p_1}\right)^{\frac{\gamma-1}{\gamma}}\right\}} \tag{5.5}$$

The performance of nozzles is illustrated in Figure 5.4.

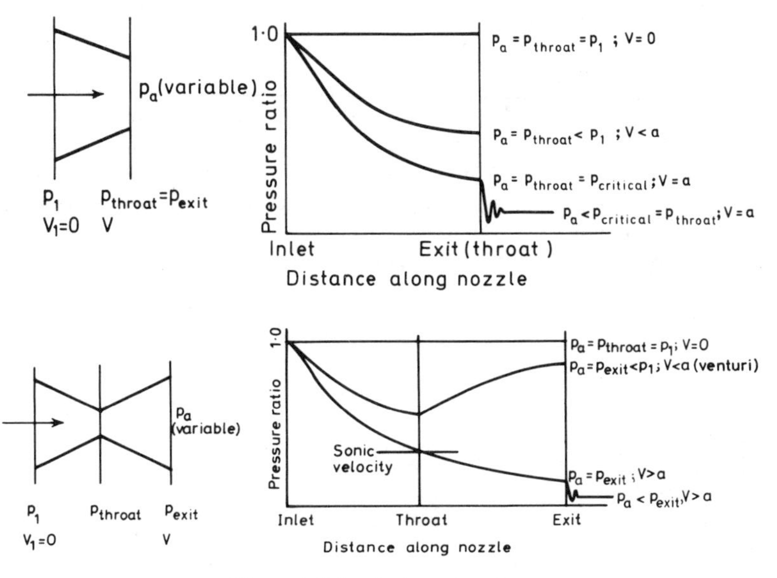

Figure 5.4 Nozzle characteristics

5.4 Steam nozzles

The critical pressure ratio and sonic velocity relations above may also be used for steam with the index k substituted for γ. Values of k are given below.

Substance	k	*Critical pressure ratio*
Superheated steam	1.3	0.546
Saturated steam	1.135	0.577
Air	1.4 (γ)	0.528

The performance of steam nozzles is complicated by the occurrence of *supersaturation* when the steam fails to condense during the expansion according to the equilibrium laws. Instead, steam initially superheated tends for a limited part of the nominally wet expansion to continue to behave as if it were superheated, resulting in higher mass flow rates than expected. If supersaturation is anticipated the specific volume should be determined in the wet region by

$$pv^k = \text{constant, with } k = 1.3$$

rather than by using the normal ($v = v_f + xv_{fg}$) relation. Such flow is termed *metastable* and only persists for a limited range of expansion.

WORKED EXAMPLES

Example 5.1 Stagnation properties

Write a program to determine the stagnation temperature ratio and the stagnation pressure ratio for a gas with γ = 1.4 flowing at Mach numbers of 1, 2, 3, 4, 5 and 6. Then determine the stagnation point temperature on an aircraft flying at Mach 3.5 at an altitude where the ambient temperature is –50°C.

The necessary reading for this example is in section 5.1.

```
LIST
 10  PRINT "EXAMPLE 5.1"
 20  PRINT
 30  DIM MA[6],RT[6],RP[6]
 40  FOR N=0 TO 5
 50   READ MA[N]
 60  NEXT N
 70  G=1.4
 80  I=G/(G-1)
 90  FOR N=0 TO 5
 100   RT[N]=1+(((G-1)/2)*(MA[N]↑2))
 110   RP[N]=RT[N]↑I
 120  NEXT N
 130  PRINT "MACH NUMBER TEMPERATURE RATIO   PRESSURE RATIO"
 140  FOR N=0 TO 5
 150   PRINT MA[N],RT[N],RP[N]
 160  NEXT N
 165  REM INTERPOLATE TO FIND TEMPERATURE RATIO AT MA=3.5
 170  C=0
 180  FOR J=0 TO 5
 190   Y=1
 200   FOR N=0 TO 5
 210    IF N=J THEN GOTO 230
 220    Y=Y*(3.5-MA[N])/(MA[J]-MA[N])
 230   NEXT N
 240   C=C+Y*RT[J]
 250  NEXT J
 260  T=C*(273-50)
 270  T=T-273
 280  PRINT "STAGNATION TEMPERATURE=";T;"C"
 290  DATA 1,2,3,4,5,6
 300  END
RUN
EXAMPLE 5.1
MACH NUMBER TEMPERATURE RATIO   PRESSURE RATIO
 1              1.2             1.89292916
 2              1.8             7.824449
 3              2.8             36.7327218
 4              4.2             151.835218
 5              6               529.089784
 6              8.2             1578.87767
STAGNATION TEMPERATURE= 496.35C
STOP AT 300
```

Program nomenclature

MA Mach number
RT temperature ratio
RP pressure ratio
G gamma
T temperature

Program notes

(1) Lines 40 to 60 read in Mach number data.
(2) Lines 70 to 120 find the required ratios.
(3) Lines 130 to 150 display the results.
(4) Lines 170 to 270 interpolate in the result table to find the required stagnation point temperature (this could also be done by putting Ma = 3.5 in the main program).
(5) It is informative to plot the stagnation temperature and stagnation pressure ratios against Mach number.

Example 5.2 Nozzle flow

Design a circular nozzle to expand jet engine exhaust gases from 110 kN/m^2, 800°C to 30 kN/m^2 in a reversible adiabatic process. For the gases $\gamma = 1.33$ and $R = 0.285$ kJ/kg K. The nozzle inlet velocity may be neglected and the gas mass flow rate is 30 kg/s. The design should give pressure ratio, velocity and nozzle diameter at any section. The pressure should be decreased in 10 kN/m^2 steps. Use the results to find the pressure ratio at the point of minimum area and compare this with the critical pressure ratio.

The necessary reading for this example is in section 5.3.

```
LIST
 10  PRINT "EXAMPLE 5.2"
 20  PRINT
 30  DIM P[8],RP[8],V[8],T[8],D[8],A[8]
 40  FOR N=0 TO 7
 50   READ P[N]
 60  NEXT N
 70  P=110
 80  T=1073
 90  R=0.285
 100  G=1.33
 110  M=30
 120  I=(G-1)/G
 130  FOR N=0 TO 7
 140   RP[N]=P[N]/P
 150   V[N]=((1000*2*G*R*T*(1-((RP[N])↑I)))/(G-1))↑0.5
 160   T[N]=T*(RP[N]↑I)
 170   A[N]=(M*R*T[N])/(P[N]*V[N])
 180   D[N]=((4*A[N])/3.142)↑0.5
 190  NEXT N
 200  PRINT "PRESSURE RATIO  VELOCITY(M/S)     DIAMETER(M)"
 210  FOR N=0 TO 7
 220   PRINT RP[N],V[N],D[N]
 230  NEXT N
 240  REM IDENTIFY THE MINIMUM DIAMETER AND CHECK PRESSURE RATIO
 250  REM INLET VELOCITY HAS BEEN NEGLECTED
 260  DATA 100,90,80,70,60,50,40,30
 270  END

RUN
EXAMPLE 5.2

PRESSURE RATIO  VELOCITY(M/S)     DIAMETER(M)
 0.9090909       240.018326        0.689370096
 0.818181818     346.015668        0.597349938
```

```
0.727272727     432.750939     0.558326148
0.636363636     511.372029     0.540057468
0.545454545     586.675539     0.534290480
0.454545455     661.808068     0.538738216
0.363636364     739.707072     0.554174148
0.272727273     824.1891       0.584968072

STOP AT 270
```

Program nomenclature

P	inlet pressure
T	inlet temperature
P(N)	exit pressure
RP	pressure ratio
V	velocity
G	gamma
R	specific gas constant
D	diameter
A	cross-sectional area

Program notes

(1) Lines 40 to 60 read in pressure data.
(2) Lines 70 to 190 find the required parameters from equation 5.5 for velocity and the continuity equation for area.
(3) Lines 200 to 230 display the results.
(4) Line 240 sets a problem which can be solved by plotting or putting eight new values of pressure into the program. These data values should be grouped around the smallest diameter already identified.

PROBLEMS

(5.1) Figure 5.1 shows a pitot static tube which when pointed directly into a fluid flow will record the difference between stagnation and static pressure. In gas flows for which this pressure difference is small it is usually assumed that density is constant and

$$V = \sqrt{\frac{2(p_t - p)}{\rho}}$$

Velocity profiles and temperature profiles may be measured in ducts by traversing a pitot static tube and a similarly shaped total head temperature probe. The flow rate can be determined by considering the duct in increments of area in which the velocity is known.

Write a program to determine the volume flow rate and mean velocity of a constant temperature fluid flowing in a circular duct in which a pitot static tube is traversed across a radius. Evaluate the program with the data below. The relevant relations are obtained by considering the duct in circular annuli of mean radius r and width Δr in which the velocity is V_r, then

$$\text{Volume flow rate} = \Sigma\ 2\ \pi\ r.\ \Delta r.\ V_r$$

and

$$V_{\text{mean}} = \frac{\text{flow rate}}{\text{area}}$$

Data for air flow in a circular duct of 0.36 m. diameter are given in the table.

Radius (m)	*Pitot static reading (mm air)*
0	58.35
0.047	50.97
0.60	46.97
0.098	42.31
0.114	38.58
0.127	34.28
0.139	31.01
0.150	27.16
0.161	20.88
0.170	15.42
0.180	0

If the gas stream had a known temperature profile how would the program need to be modified to give the volume flow rate, mass flow rate and enthalpy flux?

(5.2) Write a program to design a nozzle to expand steam in equilibrium flow in a reversible adiabatic process from an entry state of 2 MN/m^2, 600°C to an exit pressure of 0.05 MN/m^2. The nozzle inlet velocity may be neglected and the steam mass flow rate is 1.5 kg/s. The design should give pressure ratio, velocity and area at any cross-section. Identify the throat velocity of the nozzle and deduce a value of the index k for isentropic expansion of the steam in a relation of the form pv^k = constant, given that sonic velocity $a = \sqrt{kpv}$.

(5.3) The nozzle of a jet engine may be convergent or convergent-divergent. Write a program to determine the thrust per unit mass flow rate of air entering the engine:

(a) if the nozzle is convergent; and

(b) if the nozzle is convergent-divergent and operating at the design condition so that the nozzle exit pressure is equal to the ambient pressure.

The program should include an allowance for nozzle efficiency (section 2.8) and should be evaluated for the data below.

Nozzle thrust F (neglecting fuel mass) is given by

$$F = \dot{m}_{\text{air}} (V_{\text{jet}} - V_{\text{aircraft}}) + A_{\text{exit}} (p_{\text{exit}} - p_{\text{ambient}})$$

Data

Nozzle inlet stagnation conditions 120 kN/m^2, 850°C
Ambient pressure 35 kN/m^2
Aircraft velocity 180 m/s
Nozzle isentropic efficiency 0.88
$\gamma = 1.3$ and $C_p = 1.15$ kJ/kg K

(5.4) Write a program to design a circular nozzle to expand refrigerant 12 in equilibrium flow in a reversible adiabatic process from an entry state of 50°C, dry saturated to an exit temperature of –20°C. The nozzle inlet velocity may be neglected and the refrigerant mass flow rate is 0.5 kg/s. The design should give pressure ratio, velocity and nozzle diameter together with a value for the index k (in a relation of the form pv^k = constant, representing the expansion process) at any cross-section.

Chapter 6

Applications, plant, etc.

ESSENTIAL THEORY

6.1 Summary

In the previous four sections the principles for the solution of problems in thermodynamics have been discussed and demonstrated. In this chapter, these principles are applied to various situations including

Example 6.1	Steam plant feed water heating
Example 6.2	Combined heat and power plant
Example 6.3	Components for gas turbine plant
Example 6.4	Compression ignition engine indicator diagrams
Example 6.5	Refrigeration plant with flash chamber
Example 6.6	Solar panel with heat pump water heating
Example 6.7	Multi-stage reciprocating air compressors
Example 6.8	Air conditioning plant and load
Example 6.9	Spark ignition engine performance

Although the problems are approached through BASIC there will, of necessity, be some extra background theory given with some problems.

6.2 Bibliography

1. Bacon, D.H., *Engineering Thermodynamics,* Butterworths, London (1972).
2. Rogers, G.F.C. and Mayhew, Y.R., *Engineering Thermodynamics, Work and Heat Transfer,* Longmans, London (1967)
3. Eastop, T.D. and McConkey, A., *Applied Thermodynamics for Engineering Technologists,* Longmans, London (1970).

WORKED EXAMPLES

Example 6.1 Steam plant feed water heating

In problem 4.4 it was found that steam plant cycle efficiency is improved by increasing superheat temperature and by increasing steam generator pressure. Another method of improving efficiency is feed water heating (see problem 6.1 for a further method).

A steam plant with steam generator pressure and temperature 5 MN/m^2, 520°C expands steam reversibly and adiabatically to 0.004 MN/m^2. The plant is fitted with an adiabatic feed water heater in which steam bled from the turbine is condensed and as it condenses heats the feed water to the saturation temperature corresponding to the bleed pressure (Figure 6.1). The condensed bleed steam is pumped to the feed

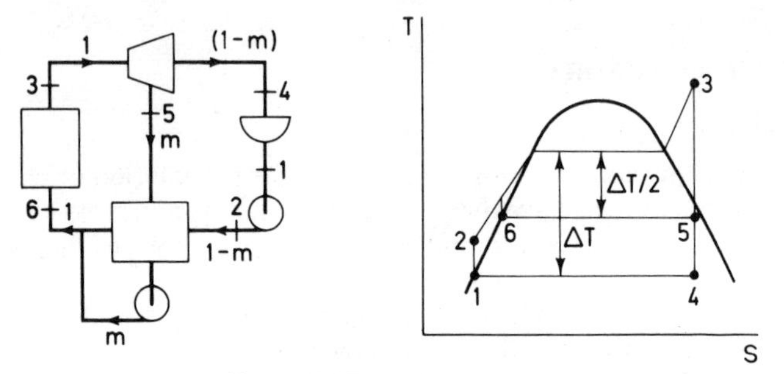

Figure 6.1 Feed water heating

water pressure and mixed with the heated feed. Show that the efficiency of the plant cycle is increased by the use of the feed heater and that the improvement in efficiency is maximised if the feed heater bleed pressure is chosen to give approximately equal increments in feed water temperature rise. All pump work may be neglected.

The amount of steam bled from the turbine may be found by application of the steady flow energy equation to the feed heater.

$$mh_5 + (1 - m)h_2 = h_6$$

The cycle efficiency is given by

$$\eta = \frac{(h_3 - h_5) + (1 - m)(h_5 - h_4)}{(h_3 - h_6)} \tag{6.1}$$

The necessary reading for this example is in section 4.8.

```
LIST
 10  PRINT "EXAMPLE .6.1"
 20  PRINT
 30  DIM P[120],T[120],H[120],S[120],U[120],V[120]
 40  FOR N=0 TO 119
 50   READ P[N],T[N],H[N],S[N],U[N],V[N]
 60  NEXT N
 70  N=0
 80  H1=H[0]
 90  N=117
 100  H3=H[117]
 110  S3=S[117]
 120  REM INTERPOLATE FOR H AT END OF EXPANSION
```

```
130  SF=S[0]
140  SG=S[10]
150  DF=(S3-SF)/(SG-SF)
160  HF=H[0]
170  HG=H[10]
180  H4=HF+DF*(HG-HF)
190  PRINT "H4=";H4;"KJ/KG"
200  PRINT "H3=";H3;"KJ/KG"
210  PRINT "H1=";H1;"KJ/KG"
220  EFF=100*(H3-H4)/(H3-H1)
230  PRINT "CYCLE EFFICIENCY WITHOUT A FEED HEATER ";EFF;"%"
240  PRINT
250  PRINT "INTERPOLATE FOR H AT 8 BLEED PRESSURES"
260  PRINT
270  PRINT "INPUT W AND M ,THEN F FOR 5 PRESSURES WHEN BLEED IS WET STEAM"
280  PRINT "INPUT D,E AND F THEN F FOR THE REMAINING 3 PRESSURES WHICH ARE"
285  PRINT "SUPERHEATED BLEED"
290  REM W AND M ARE 1,11;2,12;3,13;4;14;5,15,F IS 1,2,3,4,5
300  PRINT
310  REM D,E AND F ARE  80,89,6;90,99,7;100,109,8,F IS 6,7,8
320  DIM PB[8],M[8],E[8]
330  FOR Y=0 TO 7
335   IF Y<=4 THEN GOTO 560
340   INPUT D,E,F
350   C=0
360   FOR J=D TO E
370    Z=1
380    FOR N=D TO E
390     IF N=J THEN GOTO 410
400     Z=Z*(S3-S[N])/(S[J]-S[N])
410    NEXT N
420    C=C+Z*H[J]
430   NEXT J
440   H5=C
450   REM FIND H6(=HF) AT 8 BLEED PRESSURES WITH F=1 TO 8
455   INPUT F
460   N=F
470   PB[Y]=P[N]
480   H6=H[N]
490   REM FIND BLEED MASS
500   M[Y]=(H6-H1)/(H5-H1)
510   REM FIND EFFICIENCY
520   E[Y]=100*((H3-H5)+((1-M[Y])*(H5-H4)))/(H3-H6)
540  NEXT Y
550  GOTO 660
560  REM FOR WET STEAM  NEED HF AND HG
570  REM INPUT W AND M ARE 1,11;2,12;3,13;4,14;5,15
580  INPUT W,M
590  HF=H[W]
600  HG=H[M]
610  SF=S[W]
620  SG=S[M]
630  DF=(S3-SF)/(SG-SF)
640  H5=HF+DF*(HG-HF)
650  GOTO 450
660  PRINT "BLEED PRESSURE    BLEED MASS    EFFICIENCY   "
665  PRINT "     MN/M↑2            KG/KG          %"
670  PRINT
680  FOR Y=0 TO 7
690   PRINT PB[Y],M[Y],E[Y]
700  NEXT Y
710  END


RUN
EXAMPLE 6.1

H4= 2119.42438KJ/KG
H3= 3480KJ/KG
H1= 121.4KJ/KG
CYCLE EFFICIENCY WITHOUT A FEED HEATER  40.5102014%
```

```
INTERPOLATE FOR H AT 8 BLEED PRESSURES

INPUT W AND M ,THEN F FOR 5 PRESSURES WHEN BLEED IS WET STEAM
INPUT D,E AND F THEN F FOR THE REMAINING 3 PRESSURES WHICH ARE SUPERHEATED BLEED

? 1? 11
? 1
? 2? 12
? 2
? 3? 13
? 3
? 4? 14
? 4
? 5? 15
? 5
? 80? 89? 6
? 6
? 90? 99? 7
? 7
? 100? 109? 8
? 8

BLEED PRESSURE     BLEED MASS        EFFICIENCY

 0.01             0.0334007752      41.2660783
 0.02             0.0592016465      41.7767371
 0.05             0.0942205034      42.3530129
 0.1              0.121690559       42.6976760
 0.2              0.150366512       42.9439579
 0.5              0.19051           43.0479683
 1                0.222654357       42.8440589
 2                0.257250486       42.2870174

STOP AT 710
```

Program nomenclature

SF	saturated liquid specific entropy
SG	saturated steam specific entropy
DF	dryness fraction
HF	saturated liquid specific enthalpy
HG	saturated steam specific enthalpy
EFF	cycle efficiency with no feed heater
PB(N)	bleed pressure
M(N)	bleed mass
E(N)	cycle efficiency with feed heater

Program notes

(1) Lines 70 to 230 find the efficiency with no feed heater.
(2) Lines 270 to 310 select line numbers in the steam data block to find the bleed enthalpy for steam which is superheated or wet in the turbine expansion.
(3) Line 335 dictates the superheated or wet route.
(4) Lines 340 to 540 are the superheated route.
(5) Lines 560 to 650, then 450 to 540, are the wet route.

(6) Lines 660 to 700 arrange the results in a table.
(7) The maximum efficiency would occur at a pressure corresponding to the saturation pressure at

$$\tfrac{1}{2}(T_{sat} \text{ at } 5 \text{ MN/m}^2 + T_{sat} \text{ at } 0.004 \text{ MN/m}^2)$$
$$= \tfrac{1}{2}(263.9 + 29) = 146.45°\text{C}$$

That is at about 0.42 MN/m^2 if there are equal increments in feed water temperature rise. The results show that this is approximately true.
(8) The units of bleed mass 'kg/kg' mean kg of steam bled per kg of steam flow through the steam generator.

Example 6.2 Combined heat and power plant

Many factories use shaft power and process or space heating. Normally the shaft power is taken from electric motors using electricity from the grid and the heating is provided by a fossil fuel fired boiler plant. In the back pressure turbine plant and the pass out turbine plant. In

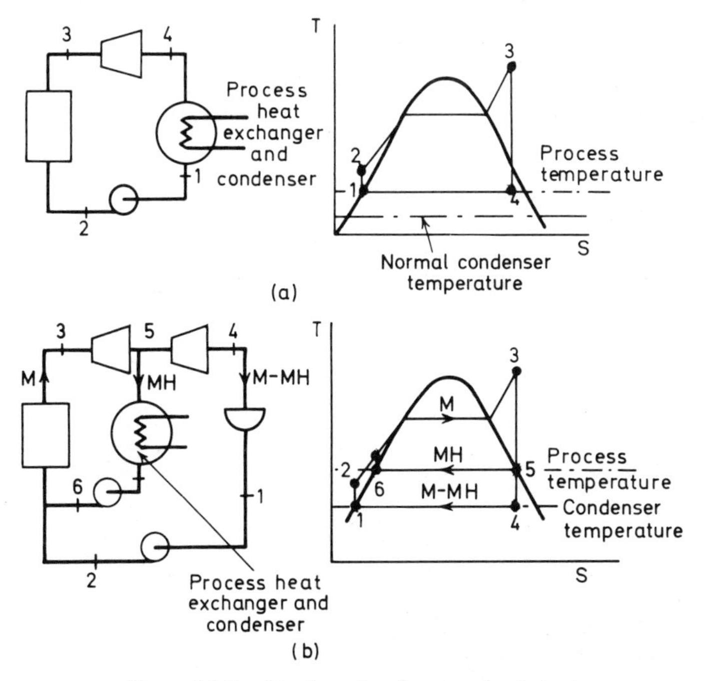

Figure 6.2 Combined work and process heat plant

a combined heat and power steam plant these energy supplies are both provided by a single plant. Figure 6.2 shows two types, the back pressure plant, condensation takes place at the temperature at which heat is required (with due allowance for heat exchanger temperature differences). Thus the turbine shaft work is reduced and the efficiency is given by

$$\eta = \frac{\text{work output} + \text{heat output}}{\text{energy input in fuel}} \tag{6.2}$$

In the back pressure turbine plant, the ratio of shaft power to heat is fixed. In the pass out turbine plant, steam is bled from the turbine at the appropriate pressure so that it supplies heat at the required temperature as it condenses. The remaining steam is expanded to the normal condenser pressure. This allows some flexibility in the ratio of shaft power to heat and the efficiency is again given by

$$\eta = \frac{\text{work output} + \text{heat output}}{\text{energy input in fuel}}$$

Write a program for the pass out plant to determine the effect of (a) varying the temperature of the process, and (b) varying the ratio of shaft work to process heating on the flow rates and efficiency of the plant.

All expansions are reversible and adiabatic and the heat exchanger and steam generator efficiencies are both 80 per cent.

Data

For (a): Shaft work 1000 kW
Process heat 1000 kW
Process temperatures are 60.1, 81.3, 99.6, 120.2 and 151.8°C (exact lines in data block).

For (b): Process temperature 99.6°C
Shaft work 1000 kW
Process heats are 500, 1000, 1500 and 2000 kW.

Steam conditions:
Condenser pressure 0.004 MN/m^2
Steam generator exit 2 MN/m^2, 300°C.

```
LIST
 10  PRINT "EXAMPLE 6.2"
 20  PRINT
 30  DIM P[120],T[120],H[120],S[120],U[120],V[120]
 40  FOR N=0 TO 119
 50   READ P[N],T[N],H[N],S[N],U[N],V[N]
 60  NEXT N
 70  H3=H[103]
 80  S3=S[103]
 90  H1=H[0]
```

```
100  REM INTERPOLATE FOR H AT END OF EXPANSION
110  SF=S[0]
120  SG=S[10]
130  DF=(S3-SF)/(SG-SF)
140  HF=H[0]
150  HG=H[10]
160  H4=HF+DF*(HG-HF)
170  PRINT "ARE YOU VARYING PROCESS TEMP,TYPE 1,OR VARYING WORK"
175  PRINT "TO HEAT RATIO,TYPE 2?"
180  INPUT X
190  IF X=2 THEN GOTO 500
200  PRINT "VARYING PROCESS TEMP WITH FIXED WORK TO HEAT RATIO"
210  PRINT "WHAT IS SHAFT WORK NEED?"
220  INPUT SW
230  PRINT "WHAT IS THE HEAT NEED?"
240  INPUT Q
250  REM    INTERPOLATE FOR H5 AT 5 PROCESS TEMPERATURES
260  REM DATA REQUIRED FOR W AND M ARE RELEVANT VALUES OF N FOR F AND G
265  REM DATA AT EACH PROCESS TEMPERATURE
270  PRINT "PROCESS       BLEED MASS      TOTAL MASS        EFFICIENCY"
280  PRINT "TEMP C            KG/S            KG/S                 %   "
290  PRINT
300  PRINT
310  DIM W[5],M[5]
320  FOR Y=0 TO 4
330   READ W[Y],M[Y]
340   HF=H[W[Y]]
350   HG=H[M[Y]]
360   SF=S[W[Y]]
370   SG=S[M[Y]]
380   DF=(S3-SF)/(SG-SF)
390   H5=HF+DF*(HG-HF)
400   H6=H[W[Y]]
410   TB=T[W[Y]]
420   MH=Q/(0.8*(H5-H6))
430   MT=(SW+MH*(H5-H4))/(H3-H4)
440   E=0.8*(SW+Q)*100/((MT-MH)*(H3-H1)+MH*(H3-H6))
450   PRINT TB,MH,MT,E
460  NEXT Y
470  PRINT
480  PRINT
490  GOTO 170
500  PRINT "VARYING WORK TO HEAT RATIO WITH PROCESS TEMP 99.6 C"
510  H6=H[4]
520  SF=S[4]
530  SG=S[14]
540  HF=H[4]
550  HG=H[14]
560  DF=(S3-SF)/(SG-SF)
570  H5=HF+DF*(HG-HF)
580  REM DATA REQUIRED FOR WORK(SW) AND HEAT(Q) FOR 4 CASES
590  PRINT "SHAFT WORK    PROCESS HEAT      BLEED MASS"
600  PRINT "     KW              KW             KG/S "
610  PRINT
620  DIM SW[4],Q[4],MT[4],E[4]
630  FOR Z=0 TO 3
640   READ SW[Z],Q[Z]
650   MH=Q[Z]/(0.8*(H5-H6))
660   MT[Z]=(SW[Z]+MH*(H5-H4))/(H3-H4)
670   E[Z]=0.8*100*(SW[Z]+Q[Z])/((MT[Z]-MH)*(H3-H1)+MH*(H3-H6))
680   PRINT SW[Z],Q[Z],MH
685  NEXT Z
690  PRINT "TOTAL MASS         EFFICIENCY"
700  PRINT "  KG/S                  %"
710  PRINT
715  FOR Z=0 TO 3
720   PRINT MT[Z],E[Z]
730  NEXT Z
740  END
1010  DATA 0.004,29,121.4,0.423,121.4,0.001
1020  DATA 0.01,45.8,191.8,0.649,191.8,0.00101
1030  DATA 0.02,60.1,251.5,0.832,251.4,0.00102
      ETC
```

```
2210LIST
 2210  DATA 2,12
 2220  DATA 3,13
 2230  DATA 4,14
 2240  DATA 5,15
 2250  DATA 6,16
 2260  DATA 1000,500
 2270  DATA 1000,1000
 2280  DATA 1000,1500
 2290  DATA 1000,2000

RUN
EXAMPLE 6.2

ARE YOU VARYING PROCESS TEMP,TYPE 1;OR VARYING WORK
TO HEAT RATIO,TYPE,2?
? 1
VARYING PROCESS TEMP WITH FIXED WORK TO HEAT RATIO
WHAT IS SHAFT WORK NEED?
? 1000
WHAT IS THE HEAT NEED?
? 1000

PROCESS        BLEED MASS       TOTAL MASS         EFFICIENCY
TEMP C           KG/S             KG/S                 %


 60.1            0.631686433      1.13680417       49.7104031
 81.3            0.620972529      1.21234636       47.2806012
 99.6            0.613357682      1.27330389       45.5120826
 120.2           0.606450860      1.33829980       43.7944054
 151.8           0.599067354      1.43206037       41.5865880


ARE YOU VARYING PROCESS TEMP,TYPE 1;OR VARYING WORK TO HEAT RATIO,TYPE,2?
? 2
VARYING WORK TO HEAT RATIO WITH PROCESS TEMP 99.6 C
SHAFT WORK    PROCESS HEAT      BLEED MASS
    KW            KW               KG/S

 1000           500             0.306678841
 1000           1000            0.613357682
 1000           1500            0.920036524
 1000           2000            1.22671536
TOTAL MASS          EFFICIENCY
   KG/S                 %

1.14378026        37.1485657
1.27330389        45.5120826
1.40282751        52.6201250
1.53235114        58.7356512

STOP AT 740
```

Program nomenclature

SF saturated liquid specific entropy
SG saturated steam specific entropy
DF dryness fraction
HF saturated liquid specific enthalpy
HG saturated steam specific enthalpy
TB bleed temperature
MH process heat mass flow rate
MT total mass flow rate

E cycle efficiency
SW shaft power
Q process heat rate

Program notes

(1) There are two programs here, one to vary process temperature and one to vary work-to-heat ratio. The separation line is 490.
(2) In the first case the program evaluates the constant specific enthalpy values and then those that vary with bleed pressure. These are used to find the relevant flow rates and efficiency. In the second case all the enthalpy values are fixed and only the ratio of work to heat is varied.
(3) The results show that, with varying process temperature, efficiency falls as the process temperature rises, which is due to the extra mass flow needed to supply the work. With varying work-to-heat ratio, the efficiency rises as the proportion of heat rises since the efficiency of power production is less than the heating efficiency.
(4) It will be noted that in order to avoid excess interpolation all the process temperatures are chosen to have wet steam at the bleed point.

Example 6.3 Components for gas turbine plant

Most practical gas turbine plant does not use the closed Joule cycle but dispenses with the cooler and replaces the heater with a combustion chamber in which fuel is burned directly in the compressed air. The products of combustion are exhausted to atmosphere after expansion in the turbine and a continuous supply of fresh air is drawn into the compressor.

Applications can be divided in three ways:
(a) stationary plant for pumping or power generation running at constant speed;
(b) low-speed traction such as automobiles, trains and ships which need speed variation; and
(c) aircraft propulsion.

For (a) and (b) a heat exchanger (between exhaust gas and compressed air) is usually fitted to improve efficiency but not for (c) since a heat exchanger is too bulky for aircraft applications. For aircraft jet propulsion an intake diffuser and an exhaust nozzle are required.

For propeller-driven aircraft and ships, and for vehicles, the turbine is split into two parts, one of which drives the compressor and the other, which is on a separate shaft and is able to rotate at any speed, drives the load.

The table below shows the possible components needed to form most types of plant together with equations for their analysis.

Write a program for each component so that they may be combined to form a program for any particular plant (see problems 6.4 and 6.5).

The necessary reading for this example is in section 4.9.

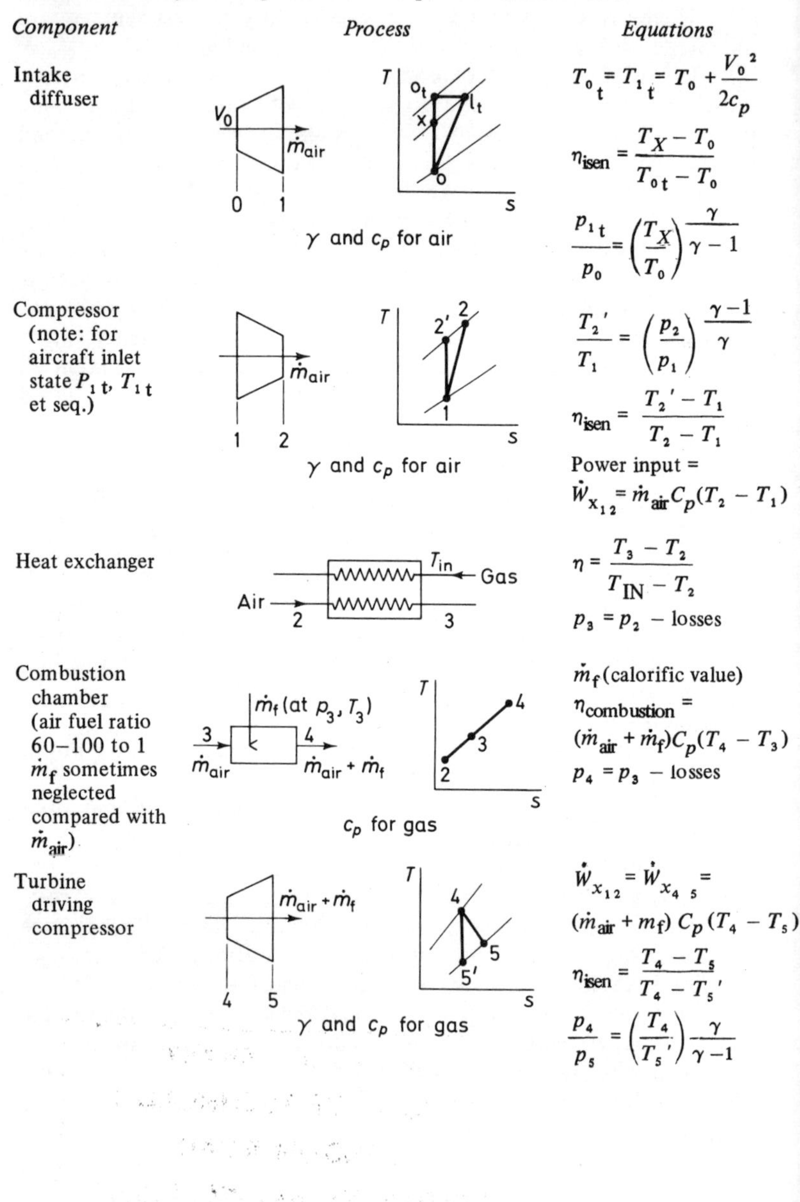

Component	*Process*	*Equations*
Intake diffuser	γ and c_p for air	$T_{0_t} = T_{1_t} = T_0 + \dfrac{V_0^2}{2c_p}$ $\eta_{isen} = \dfrac{T_X - T_0}{T_{0t} - T_0}$ $\dfrac{p_{1t}}{p_0} = \left(\dfrac{T_X}{T_0}\right)^{\frac{\gamma}{\gamma - 1}}$
Compressor (note: for aircraft inlet state $P_{1\,t}$, $T_{1\,t}$ et seq.)	γ and c_p for air	$\dfrac{T_2{}'}{T_1} = \left(\dfrac{p_2}{p_1}\right)^{\frac{\gamma - 1}{\gamma}}$ $\eta_{isen} = \dfrac{T_2{}' - T_1}{T_2 - T_1}$ Power input = $\dot{W}_{x_{12}} = \dot{m}_{air} C_p (T_2 - T_1)$
Heat exchanger		$\eta = \dfrac{T_3 - T_2}{T_{IN} - T_2}$ $p_3 = p_2$ – losses
Combustion chamber (air fuel ratio 60–100 to 1 $\dot{m}_f$ sometimes neglected compared with $\dot{m}_{air}$)	c_p for gas	$\dot{m}_f$(calorific value) $\eta_{combustion}$ = $(\dot{m}_{air} + \dot{m}_f) C_p (T_4 - T_3)$ $p_4 = p_3$ – losses
Turbine driving compressor	γ and c_p for gas	$\dot{W}_{x_{12}} = \dot{W}_{x_{4\,5}} = (\dot{m}_{air} + m_f)\, C_p (T_4 - T_5)$ $\eta_{isen} = \dfrac{T_4 - T_5}{T_4 - T_5{}'}$ $\dfrac{p_4}{p_5} = \left(\dfrac{T_4}{T_5{}'}\right)^{\frac{\gamma}{\gamma - 1}}$

Component	*Process*	*Equations*
Turbine driving load/or load and compressor)	γ and c_p for gas; p_6 is atmospheric	$\frac{T_5}{T_6'} = \left(\frac{p_5}{p_6}\right)^{\frac{\gamma-1}{\gamma}}$

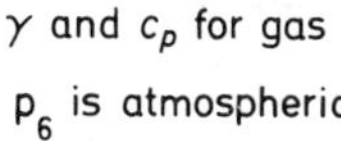

$$\eta_{isen} = \frac{T_5 - T_6}{T_5 - T_6'}$$

Power output =

$$\dot{W}_{x_{56}} = (\dot{m}_{air} + \dot{m}_f)' C_p(T_5 - T_6)$$

but if driving load and compressor net power =

$$\dot{W}_{x_{56}} - \dot{W}_{x_{12}}$$

Component	*Process*	*Equations*
Exhaust nozzle (note: for aircraft total head properties at nozzle inlet p_{6_t}, T_{6_t})	γ and c_p for gas	$\frac{T_{6t}}{T_7'} = \left(\frac{p_{6t}}{p_7}\right)^{\frac{\gamma-1}{\gamma}}$

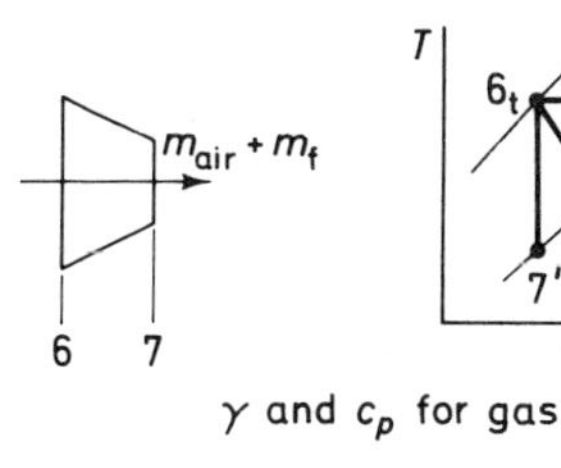

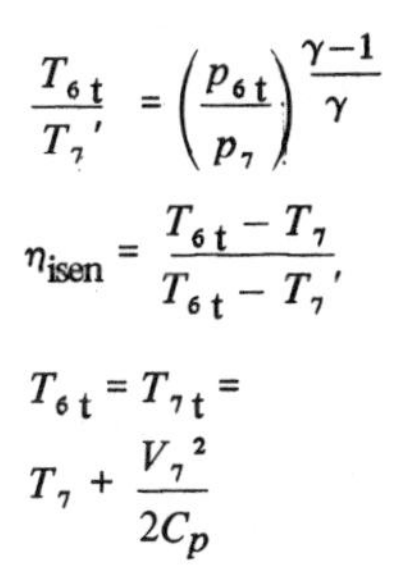

$$\eta_{isen} = \frac{T_{6t} - T_7}{T_{6t} - T_7'}$$

$$T_{6t} = T_{7t} = T_7 + \frac{V_7^2}{2C_p}$$

$$\text{Thrust } F = (\dot{m}_{air} + \dot{m}_f) V_7 - \dot{m}_{air} V_0$$

If a convergent divergent nozzle fully expanding to atmospheric pressure.

$$\text{Thrust } F = (\dot{m}_{air} + \dot{m}_f) V_7 - \dot{m}_{air} V_0 + A_{exit} (p_7 - p_0)$$

if not expanding to atmospheric or if convergent, in which case normal choking critical pressure will give the value of p_7

```
LIST
 10  PRINT "EXAMPLE 6.3"
 20  PRINT
 30  PRINT "INTAKE DIFFUSER"
 40  PRINT "DATA INPUT,PIN,TIN,VELOCITY IN,EFFICIENCY,CP,GAMMA"
 50  INPUT PIN,TIN,VIN,E1,CPA,GA
 60  T0=TIN+(VIN↑2)/(2000*CPA)
 70  T1=T0
 80  TX=(E1*(T0-TIN))+TIN
 90  I=GA/(GA-1)
 100  P1=PIN*((TX/TIN)↑I)
 110  PRINT
 120  PRINT "STAGNATION TEMPERATURE AT EXIT=";T1;"K"
 130  PRINT "STAGNATION PRESSURE AT EXIT=";P1;"KN/M↑2"
 140  PRINT
 150  END

RUN
EXAMPLE 6.3

INTAKE DIFFUSER
DATA INPUT,PIN,TIN,VELOCITY IN,EFFICIENCY,CP,GAMMA
? 60? 260? 190? 0.82? 1.01? 1.4

STAGNATION TEMPERATURE AT EXIT= 277.871287K
STAGNATION PRESSURE AT EXIT= 72.6938721KN/M↑2


STOP AT 150



200LIST
 200  PRINT "COMPRESSOR"
 210  PRINT "DATA INPUT,PIN,POUT,TIN,EFFICIENCY,CP,GAMMA,MASS FLOW RATE"
 220  INPUT P1,P2,T1,E2,CPA,GA,MA
 230  I=(GA-1)/GA
 240  T21=T1*((P2/P1)↑I)
 250  T2=((T21-T1)/E2)+T1
 260  WXC=MA*CPA*(T2-T1)
 270  PRINT
 280  PRINT "PRESSURE AT EXIT=";P2;"KN/M↑2"
 290  PRINT "TEMPERATURE AT EXIT=";T2;"K"
 300  PRINT "COMPRESSOR POWER=";WXC;"KW"
 310  PRINT
 320  END

RUN
COMPRESSOR
DATA INPUT,PIN,POUT,TIN,EFFICIENCY,CP,GAMMA,MASS FLOW RATE
? 72.69? 480? 277.87? 0.81? 1.01? 1.4? 20

PRESSURE AT EXIT= 480KN/M↑2
TEMPERATURE AT EXIT= 523.0892K
COMPRESSOR POWER= 4953.428KW


STOP AT 320


400LIST
 400  PRINT "HEAT EXCHANGER"
 410  PRINT "DATA INPUT,TIN,PIN,T GAS,EFFICIENCY,PRESSURE LOSS"
 420  INPUT T2,P2,TG,E3,PL
 430  T3=(E3*(TG-T2))+T2
 440  P3=P2-PL
 450  PRINT
 460  PRINT "PRESSURE AT EXIT=";P3;"KN/M↑2"
 470  PRINT "TEMPERATURE AT EXIT=";T3;"K"
 480  PRINT
 490  END
```

```
RUN
HEAT EXCHANGER
DATA INPUT,TIN,PIN,T GAS,EFFICIENCY,PRESSURE LOSS
? 491? 450? 680? 0.68? 20

PRESSURE AT EXIT= 430KN/M↑2
TEMPERATURE AT EXIT= 619.52K


STOP AT 490
```

```
600LIST
 600  PRINT "COMBUSTION CHAMBER"
 610  PRINT "DATA INPUT,TIN,PIN,PRESSURE LOSS,FUEL FLOW"
 615  PRINT "CALORIFIC VALUE,EFFICIENCY,CP GAS,AIR FLOW"
 620  INPUT T3,P3,PL,MF,CAL,E4,CPG,MA
 630  P4=P3-PL
 640  T4=((MF*CAL*E4)/((MA+MF)*CPG))+T3
 650  PRINT
 660  PRINT "PRESSURE AT EXIT=";P4;"KN/M↑2"
 670  PRINT "TEMPERATURE AT EXIT=";T4;"K"
 680  PRINT
 690  END
```

```
RUN
COMBUSTION CHAMBER
DATA INPUT,TIN,PIN,PRESSURE LOSS,FUEL FLOW
CALORIFIC VALUE,EFFICIENCY,CP GAS,AIR FLOW
? 620? 450? 20? 0.16? 45000? 0.963? 1.15? 20

PRESSURE AT EXIT= 430KN/M↑2
TEMPERATURE AT EXIT= 919.068323K


STOP AT 690
```

```
800LIST
 800  PRINT "TURBINE DRIVING COMPRESSOR"
 810  PRINT "DATA INPUT,COMPRESSOR POWER,AIR FLOW,FUEL FLOW"
 815  PRINT "CP GAS,GAMMA GAS,EFFICIENCY,PIN,TIN"
 820  INPUT WXC,MA,MF,CPG,GG,E5,P4,T4
 830  WXT=WXC
 840  T5=T4-(WXT/((MA+MF)*CPG))
 850  T51=T4-((T4-T5)/E5)
 860  I=GG/(GG-1)
 870  P5=P4/((T4/T51)↑I)
 880  PRINT
 890  PRINT "EXIT PRESSURE=";P5;"KN/M↑2"
 900  PRINT "EXIT TEMPERATURE=";T5;"K"
 910  PRINT
 920  END
```

```
RUN
TURBINE DRIVING COMPRESSOR
DATA INPUT,COMPRESSOR POWER,AIR FLOW,FUEL FLOW
CP GAS,GAMMA GAS,EFFICIENCY,PIN,TIN
? 4953? 20? 0.32? 1.15? 1.3? 0.85? 460? 1114

EXIT PRESSURE= 153.417088KN/M↑2
EXIT TEMPERATURE= 902.043478K


STOP AT 920
```

```
1000LIST
 1000  PRINT "TURBINE DRIVING LOAD OR COMPRESSOR AND LOAD"
 1010  PRINT "DATA INPUT,PIN,TIN,EFFICIENCY,GAMMA GAS,CP GAS"
 1015  PRINT "POUT,AIR FLOW,FUEL FLOW,COMPRESSOR POWER"
 1020  INPUT P5,T5,E6.GG,CPG,P6,MA,MF,WXC
 1030  I=(GG-1)/GG
 1040  T61=T5/((P5/P6)↑I)
 1050  T6=T5-(E6*(T5-T61))
 1060  WXT=(MA+MF)*CPG*(T5-T6)
 1070  PRINT "DOES TURBINE ONLY DRIVE LOAD?YES,TYPE 1;NO,TYPE 2"
 1080  INPUT X
 1090  IF X=2 THEN GOTO 1120
 1100  PRINT
 1110  PRINT "POWER OUTPUT OF PLANT=";WXT;"KW"
 1120  WX=WXT-WXC
 1130  PRINT "POWER OUTPUT OF PLANT=";WX;"KW"
 1140  PRINT
 1150  END


RUN
TURBINE DRIVING LOAD OR COMPRESSOR AND LOAD
DATA INPUT,PIN,TIN,EFFICIENCY,GAMMA GAS,CP GAS
POUT,AIR FLOW,FUEL FLOW,COMPRESSOR POWER
? 480? 919? 0.85? 1.3? 1.15? 100? 20? 0.16? 4020
DOES TURBINE ONLY DRIVE LOAD?YES,TYPE 1;NO,TYPE 2
? 2
POWER OUTPUT OF PLANT= 1480.23380KW


STOP AT 1150


1200LIST
 1200  PRINT "EXHAUST NOZZLE"
 1210  PRINT "DATA INPUT,PIN(STAG),TIN(STAG),PAMBIENT,EFFICIENCY,AIRCRAFT VELOCITY"
 1220  PRINT "AIR MASS,FUEL MASS,GAMMA GAS,CP GAS"
 1230  INPUT P6,T6,PA,E7,VIN,MA,MF,GG,CPG
 1240  PRINT "IS THE NOZZLE CONVERGENT OR CON-DIV?CON,TYPE 1,CON-DIV TYPE 2"
 1250  INPUT Z
 1260  IF Z=2 THEN GOTO 1450
 1270  PRINT "IS THE NOZZLE CHOKED?"
 1280  I=GG/(GG-1)
 1290  RPC=(2/(GG+1))↑I
 1300  PC=RPC*P6
 1310  I=1/I
 1320  IF PC<PA THEN GOTO 1400
 1330  T71=T6/((P6/PC)↑I)
 1340  T7=T6-E7*(T6-T71)
 1350  V7=(2000*CPG*(T6-T7))↑0.5
 1360  PRINT "INPUT NOZZLE EXIT AREA"
 1370  INPUT A
 1380  F=((MA+MF)*V7)-(MA*VIN)+A*(PC-PA)*1000
 1390  PRINT "NOZZLE CONVERGENT,CHOKED,THRUST=";F;"N"
 1395  GOTO 1570
 1400  T71=T6/((P6/PA)↑I)
 1410  T7=T6-E7*(T6-T71)
 1420  V7=(2000*CPG*(T6-T7))↑0.5
 1430  F=((MA+MF)*V7)-(MA*VIN)
 1440  PRINT "NOZZLE CONVERGENT,UNCHOKED,THRUST=";F;"N"
 1445  GOTO 1570
 1450  PRINT "ONLY FULL EXPANSION TO AMBIENT PRESSURE IS SOLVED WITH A CHOKED"
 1455  PRINT "CON-DIV NOZZLE"
 1460  I=(GG-1)/GG
 1470  T71=T6/((P6/PA)↑I)
 1480  T7=T6-E7*(T6-T71)
 1490  V7=(2000*CPG*(T6-T7))↑0.5
 1500  F=((MA+MF)*V7)-(MA*VIN)
 1510  PRINT "NOZZLE CON-DIV,CHOKED AT DESIGN CONDITION,THRUST=";F;"N"
 1520  PRINT
```

```
1530  PRINT "FIND CON-DIV EXIT AREA"
1540  RG=((GG-1)*CPG)/GG
1550  AR=((MA+MF)*RG*T7)/(V7*PA)
1560  PRINT "CON-DIV NOZZLE EXIT AREA=";AR;"M↑2"
1570  END
```

```
RUN
EXHAUST NOZZLE
DATA INPUT,PIN(STAG),TIN(STAG),PAMBIENT,EFFICIENCY,AIRCRAFT VELOCITY
AIR MASS,FUEL MASS,GAMMA GAS,CP GAS
? 170? 900? 60? 0.9? 190? 32.5? 0? 1.3? 1.15
IS THE NOZZLE CONVERGENT OR CON-DIV?CON,TYPE 1,CON-DIV TYPE 2
? 1
IS THE NOZZLE CHOKED?
INPUT NOZZLE EXIT AREA
? 0.15
NOZZLE CONVERGENT,CHOKED,THRUST= 14761.9420N

STOP AT 1570
```

Program nomenclature

Diffuser:

TO	stagnation temperature
Pl	stagnation pressure

Compressor:

T21	ideal exit temperature
WXC	compressor power

Turbine:

T51	ideal exit temperature
WXT	turbine power
T61	ideal exit temperature

Nozzle:

RPC	critical pressure ratio
PC	critical pressure
T71	ideal exit temperature
V7	exit velocity
F	thrust
AR	exit area

Program notes

(1) The program has been written as seven separate units with spare line numbers available for rearrangements which may be required for any particular problem. To make it run sequentially some END statements would need to be removed.

(2) It will not always be possible to solve a particular problem with these programs as they are written, since the data available may be different from that used as input and restructuring may be required. See problems 6.4 and 6.5.

(3) Each program section has been run with data and results are shown. The data used is not sequential to that of previous components since no plant will use all the components. In any problem these unnecessary units can be jumped by GOTO additions.
(4) The nozzle section of the program does not consider off design convergent-divergent nozzles which require more complex analysis.
(5) When dealing with aircraft plant the input data to the compressor will be stagnation values, which will continue to be used throughout the plant until exit from the nozzle when a return to static properties and velocity is required.
(6) In the test run for the nozzle the fuel flow rate is set to zero. This deals with problems in which it is stated that 'the mass flow rate of fuel may be neglected'.

Example 6.4 Compression ignition engine indicator diagrams

The processes in the working substance of a four-stroke compression ignition engine can be modelled by incrementing the crank angle and calculating the pressure and temperature after each increment from a knowledge of the pressure and temperature before the increment, the

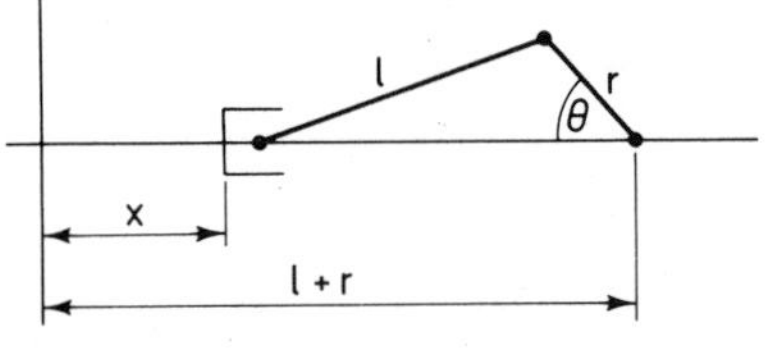

Figure 6.3

change in volume and the energy transfers. The processes can be considered in three sections: inlet valve open, both valves closed and exhaust valve open. In this example the valve-closed period is considered, for which $Q - W = \Delta U$. For a small increment in volume

$$Q - p_1 (V_2 - V_1) = mC_v (T_2 - T_1)$$

where

$$m = \left[\frac{pV}{RT}\right]_{\text{initial}}$$

This equation is solved for T_2 and then $p_2 = m R T_2/V_2$.

V_2 is obtained from the engine geometry (Figure 6.3)

$$V_2 = V_{\text{clearance}} + \frac{\pi}{4} (\text{bore})^2 . x$$

where

$$x = l + r - (r \cos\theta + \sqrt{l^2 - r^2 \sin^2\theta})$$

and $$\frac{V_{\text{maximum}}}{V_{\text{clearance}}} = \text{compression ratio}$$

Write a program for an adiabatic process to determine the pressure and temperature at any crank angle. Use the data given to evaluate the program: engine bore, 0.0762 m; stroke, 0.1111 m; connecting rod length, 0.2413 m; compression ratio, 22; speed, 1750 rev/min; valve-closed period, 216° to 496°. Assumed conditions at valve closure: 101 kN/m^2, 300 K.

Problems 6.6 and 6.7 require this program to be developed to allow for simple heat transfer through the cylinder walls and for energy release during combustion of injected fuel. A cycle could be completed by writing programs for induction and exhaust but this is not considered.

The result of this program approximates to an adiabatic process (pV^γ = constant). The addition of heat transfer approximates to a motoring curve, which may be obtained by Farnborough or electronic indicators on an actual engine. The result of the addition of heat transfer and energy release should produce a firing indicator diagram which could be integrated to give power output (ignoring the net induction and exhaust power).

```
LIST
 10  PRINT "EXAMPLE 6.4"
 20  PRINT
 30  PRINT
 40  REM ENGINE DATA
 50  CR=22
 60  RPM=1750
 70  L=0.2413
 80  S=0.1111
 90  B=0.0762
 100  REM ASSUMED INITIAL CONDITIONS
 110  PI=101
 120  TI=300
 130  REM AIR PROPERTIES
 140  R=0.287
 150  SHV=0.717
 160  A=(3.142*(B↑2))/4
 170  CV=(A*S)/(CR-1)
 180  PRINT "CLEARANCE VOLUME";CV;"M↑3"
 190  CA=216
 200  C=(CA*3.142/180)
 210  E1=(L+0.5*S-0.5*S*COS[C])
 220  E2=(0.5*S*SIN[C])
 230  E3=(L↑2)-(E2*E2)
 240  E4=E3↑0.5
 250  V1=CV+A*(E1-E4)
 260  PRINT "INITIAL VOLUME";V1;"M↑3"
 270  M=PI*V1/(R*TI)
 280  PRINT "MASS OF AIR";M;"KG"
 290  P1=PI
 300  T1=TI
 310  PRINT "CRANK ANGLE       PRESSURE          TEMPERATURE       VOLUME"
 320  PRINT "   DEG              KN/M↑2              K              M↑3"
 330  PRINT CA,P1,T1,V1
 340  FOR CA=217 TO 496
 350   C=(CA*3.142/180)
 360   E1=(L+0.5*S-0.5*S*COS[C])
```

```
370   E2=(0.5*S*SIN[C])
380   E3=(L↑2)-(E2*E2)
390   E4=E3↑0.5
400   V2=CV+A*(E1-E4)
410   DV=V2-V1
420   REM Q-W=DU,Q=0 HENCE PDV=M*SHV*(T1-T2)
430   T2=T1-(DV*P1)/(M*SHV)
440   P2=(M*R*T2)/V2
450   PRINT CA,P2,T2,V2
460   V1=V2
470   T1=T2
480   P1=P2
490  NEXT CA
500  END

RUN
EXAMPLE 6.4

CLEARANCE VOLUME 0.0000241296430M↑3
INITIAL VOLUME 0.000492527546M↑3
MASS OF AIR 0.000577761697KG
CRANK ANGLE     PRESSURE       TEMPERATURE     VOLUME
  DEG            KN/M↑2            K            M↑3
 216            101            300            0.000492527546
 217            101.6174       300.521944     0.000490386774
 218            102.257834     301.060932     0.000488189526
 219            102.921867     301.6172       0.000485936024
 220            103.610098     302.191        0.0004836265
 221            104.323149     302.782593     0.0004812612
 222            105.061673     303.392232     0.0004788404
 223            105.826350     304.020196     0.000476364358
 224            106.617893     304.666769     0.000473833362
 225            107.437044     305.332243     0.000471247718
 226            108.284582     306.016920     0.000468607742
 227            109.161319     306.721114     0.000465913767
 228            110.0681       307.445147     0.000463166141


 353            5933.35534     945.413633     0.0000264211761
 354            6129.99780     954.163226     0.0000258102969
 355            6305.34543     961.809117     0.0000252936
 356            6455.251       968.2365       0.0000248713270
 357            6575.92176     973.342247     0.0000245436743
 358            6664.12877     977.038994     0.0000243107950
 359            6717.40114     979.258961     0.0000241727975
 360            6734.18327     979.957061     0.0000241297465
 361            6713.93867     979.113118     0.0000241816618
 362            6657.18779     976.732960     0.0000243285194
 363            6565.47527     972.848275     0.00002457025
 364            6441.27036     967.515255     0.0000249067428
 365            6287.81198     960.812124     0.0000253378390
 366            6108.91566     952.835786     0.0000258633379
 367            5908.76232     943.697863     0.0000264829948


 483            114.057062     300.170139     0.000436391166
 484            112.8918       299.287068     0.000439598478
 485            111.764446     298.426582     0.000442756022
 486            110.673773     297.588253     0.000445863288
 487            109.618622     296.771670     0.000448919784
 488            108.597885     295.976431     0.000451925038
 489            107.6105       295.202148     0.0004548786
 490            106.655453     294.448447     0.000457780034
 491            105.731771     293.714962     0.000460628926
 492            104.838526     293.001341     0.000463424879
 493            103.974830     292.307243     0.000466167511
 494            103.139832     291.632338     0.000468856460
 495            102.332719     290.9763       0.000471491378
 496            101.552712     290.338835     0.000474071935

STOP AT 500
```

Program nomenclature

CR compression ratio
RPM engine speed
L connecting rod length
S stroke
B bore
PI initial pressure
TI initial temperature
R specific gas constant
SHV specific heat at constant volume
A cross-sectional area
CV clearance volume
CA crank angle
V1 volume at start of increment
P1 pressure at start of increment
T1 temperature at start of increment
M mass of air
DV increment in volume
V2 volume at end of increment
P2 pressure at end of increment
T2 temperature at end of increment

Program notes

(1) A short extract of the results is shown. The dimensions of the engine are those of a Ricardo E6/T variable compression ratio engine and if problems 6.6 and 6.7 are solved the full results may be plotted and compared with measured data. The program can be adapted to other engines by changing lines 50 to 90, 190 and 340.
(2) The program results approximate to an adiabatic process on air treated as a perfect gas. Thus for $pV^{\gamma} = c$ at a crank angle θ of 360°

$$p_{360} = p_{216}\left[\frac{V_{216}}{V_{360}}\right]^{1.4} = 6889 \text{ kN/m}^2$$

and

$$T_{360} = T_{216}\left[\frac{V_{216}}{V_{360}}\right]^{0.4} = 1002 \text{ K}$$

(3) In lines 200 to 250 and 350 to 400 the volume at crank angle θ is determined (see Figure 6.3) from

$$V_{\theta} = \text{clearance volume} + \text{area}\left[\left(L + \frac{S}{2} - \frac{S}{2}\cos\theta\right) - \sqrt{L^2 - \left(\frac{S}{2}\sin\theta\right)^2}\right]$$

where L is the connecting rod length and S is the stroke.

(4) Line 430 solves the non-flow energy equation for T_2 and line 440 uses the ideal gas rule to find p_2.
(5) Lines 460 to 480 reset the volume, temperature and pressure for the next increment of crank angle. It might be interesting to vary the amount of an increment of crank angle, for example, half a degree, two degrees, etc.
(6) To save storage the values are printed as calculated. The alternative method would involve allocating space for all the volume, pressure and temperature data with a DIM statement.

Example 6.5 Refrigeration plant with flash chamber

In order to improve the coefficient of performance of a refrigeration plant a flash chamber can be fitted in the expansion line. The flash chamber takes the wet vapour (state 5) from the throttle valve and separates the saturated liquid from the dry vapour which has no evaporative cooling potential.

The dry vapour is recycled through a second compressor and the saturated liquid is further expanded in a second throttle valve to the evaporator pressure (Figure 6.4). The flash chamber thus saves the work

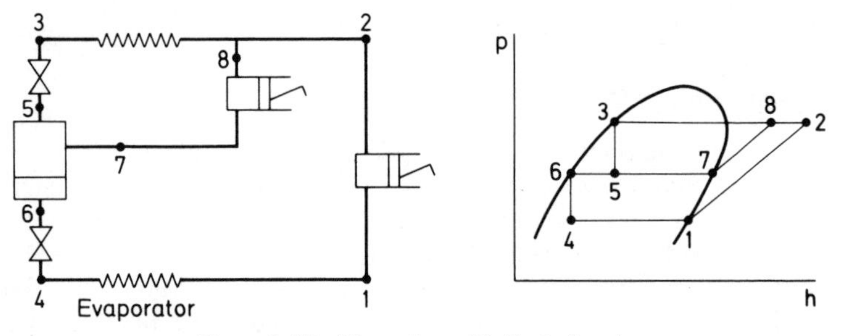

Figure 6.4 Refrigeration with flash chamber

needed to compress the dry vapour portion from evaporator pressure to flash chamber pressure and so improves the coefficient of performance. Write a program to determine the optimum pressure (giving maximum c.o.p.) at which the flash chamber should be situated for the operating conditions given below. The plant uses refrigerant 12 and the program should only use the data in the given table to avoid excessive interpolation.

If x is the dryness fraction at state 5 then the coefficient of performance is given by

$$\text{c.o.p.} = \frac{(1-x)(h_1 - h_4)}{x(h_8 - h_7) + (1-x)(h_2 - h_1)} \tag{6.3}$$

The evaporator temperature is $-5°C$ and the refrigerant leaves the evaporator as a dry vapour. Compressions are reversible and adiabatic and the condenser temperature is $35°C$. The condensate is saturated liquid.

The necessary reading for this example is in section 4.12.

```
LIST
 10  PRINT "EXAMPLE 6.5"
 20  PRINT
 30  DIM T[100],P[100],V[100],H[100],S[100]
 40  FOR N=0 TO 99
 50   READ T[N],P[N],V[N],H[N],S[N]
 60  NEXT N
 70  H1=H[29]
 80  S1=S[29]
 90  PRINT S1
 100  B=0
 110  PRINT "ENTROPY?"
 120  INPUT S1
 130  FOR J=91 TO 93
 140   X=1
 150   FOR N=91 TO 93
 160    IF N=J THEN GOTO 180
 170    X=X*(S1-S[N])/(S[J]-S[N])
 180   NEXT N
 190   B=B+X*H[J]
 200  NEXT J
 210  H2=B
 220  H3=H[17]
 230  H5=H3
 240  DIM COP[100]
 250  FOR N=10 TO 16
 260   H6=H[N]
 270   N=N+20
 280   H7=H[N]
 290   S7=S[N]
 300   DF=(H5-H6)/(H7-H6)
 310   H4=H6
 320   PRINT S7
 330   C=0
 340   PRINT "ENTROPY?"
 350   INPUT S7
 360   FOR J=91 TO 93
 370    Z=1
 380    FOR K=91 TO 93
 390     IF J=K THEN GOTO 410
 400     Z=Z*(S7-S[K])/(S[J]-S[K])
 410    NEXT K
 420    C=C+Z*H[J]
 430   NEXT J
 440   H8=C
 450   N=N-20
 460   COP[N]=(1-DF)*(H1-H4)/(DF*(H8-H7)+(1-DF)*(H2-H1))
 470  NEXT N
 480  PRINT
 490  PRINT "FLASH CHAMBER    COEFFICIENT OF"
 500  PRINT "PRESSURE         PERFORMANCE"
 510  PRINT "MN/M↑2"
 520  PRINT
 530  FOR N=10 TO 16
 540   PRINT P[N],COP[N]
 550  NEXT N
 560  END


RUN
EXAMPLE 6.5

 0.6991
ENTROPY?
```

```
? .6991
 Ø.6966
ENTROPY?
? .6966
 Ø.6942
ENTROPY?
? .6942
 Ø.6921
ENTROPY?
? .6921
 Ø.69Ø2
ENTROPY?
? .69Ø2
 Ø.6885
ENTROPY?
? .6885
 Ø.6869
ENTROPY?
? .6869
 Ø.6854
ENTROPY?
? .6854

FLASH CHAMBER   COEFFICIENT OF
PRESSURE        PERFORMANCE
MN/M↑2

 Ø.3Ø8           5.774ØØ245
 Ø.362           5.93676786
 Ø.423           6.Ø3Ø11877
 Ø.491           6.Ø6972864
 Ø.567           6.Ø3898316
 Ø.651           5.94365447
 Ø.745           5.7863Ø593

STOP AT 56Ø
```

Program nomenclature

COP coefficient of performance
DF dryness fraction
P(N) flash chamber pressure

Program notes

(1) After reading in the data tables the program calculates the specific enthalpy at evaporator exit, condenser exit and compressor exit which have fixed values (lines 70 to 230) in all cases.
(2) The flash chamber state is then varied using each of the exact lines in the data block that lie between the evaporator and condenser pressures (lines 250 to 440).
(3) The coefficient of performance is determined and tabulated in lines 450 to 550.
(4) The results show that the maximum coefficient of performance occurs when the flash chamber pressure is chosen so that the temperature difference between evaporation and condensation is divided into approximately equal parts. The coefficient of performance without a flash chamber is 5.55 so that the optimum placing of the flash

chamber gives an improvement of about 9 per cent in this ideal, reversible cycle.

Example 6.6 Solar panel with heat pump water heating

A solar panel in southern England receives an annual mean of 100 W/m^2 of solar radiation. The panel absorbs some of this radiation and loses energy by convection and re-radiation. The net energy received is transferred to the water circulated in the system. A panel efficiency is expressed by

$$\eta_{\text{panel}} = \frac{\text{energy transfer to system}}{\text{energy incident on panel}} \tag{6.4}$$

It is proposed to replace a hot water system heated by electricity costing 5p/kWh with a solar panel of 5 m^2 area having an efficiency of 20 per cent. The panel will transfer energy to water which circulates between the panel and the hot water system which uses 250 litres per day at 45°C. The cold water supply is at 10°C (Figure 6.5(a)). The balance of the water heating requirements is to be supplied by electricity costing 5p/kWh and it may be assumed that there are no losses in the system.

An alternative system utilising a heat pump is being considered (Figure 6.5(b)). The heat pump uses refrigerant 12 as a working substance which leaves the evaporator as a dry saturated vapour and leaves the condenser as a saturated liquid. The coefficient of performance of

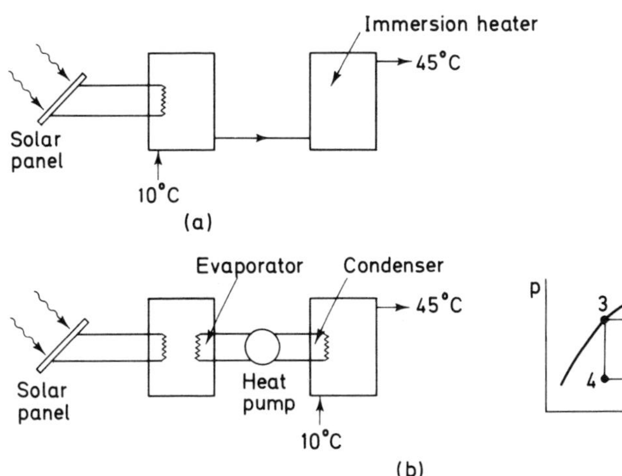

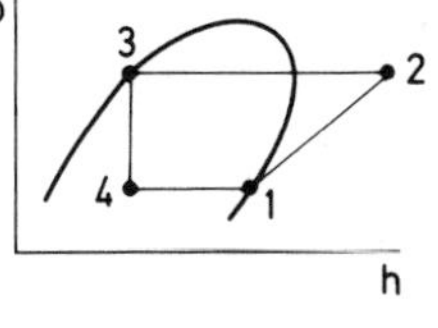

Figure 6.5

the heat pump is 30 per cent of the ideal value. The average evaporator temperature is 10°C and the condenser temperature is 50°C. The available energy input is that from the solar panel system, which will need to be increased in area to meet the demand. The heat pump compressor motor is powered by electricity costing 5p/kWh.

Write a program to find the annual cost of plain electric heating, solar panel plus electric heating and solar panel plus heat pump heating, and hence determine the cash available for installation based on 10 years' running (with no inflation, depreciation, interest, etc.) of each system. The program should allow for alterations in panel area and in flow rate from the hot water system.

The necessary reading for this example is in section 4.12.

```
LIST
 10  PRINT "EXAMPLE 6.6"
 20  PRINT
 30  PRINT "WHAT IS THE DAILY FLOW RATE IN LITRES?"
 40  INPUT F
 50  REM CALCULATE ANNUAL COST OF ELECTRIC WATER HEATING
 60  C1=(F*365*4.18*(45-10)*0.05)/3600
 70  REM CALCULATE ANNUAL COST WITH SOLAR PANEL
 80  PRINT "WHAT IS THE PANEL AREA IN M↑2?"
 90  INPUT A
 100  REM ENERGY FROM PANEL
 110  E2=(0.2*A*100*365*24*3600)/1000
 120  C2=((F*365*4.18*(45-10))-E2)*(0.05/3600)
 130  REM CALCULATE ANNUAL COST WITH A HEAT PUMP
 140  REM DETERMINE TOTAL ENERGY NEED
 150  E3=(C1*3600)/0.05
 160  PRINT "FIND COP OF HEAT PUMP"
 170  DIM T[100],P[100],V[100],H[100],S[100]
 180  FOR N=0 TO 99
 190   READ T[N],P[N],V[N],H[N],S[N]
 200  NEXT N
 220  S1=S[32]
 230  PRINT S1
 240  H3=H[19]
 250  B=0
 260  PRINT "ENTROPY?"
 270  INPUT S1
 280  FOR J=97 TO 99
 290   X=1
 300   FOR N=97 TO 99
 310    IF N=J THEN GOTO 330
 320    X=X*(S1-S[N])/(S[J]-S[N])
 330   NEXT N
 340   B=B+X*H[J]
 350  NEXT J
 360  H2=B
 370  H1=H[32]
 380  COP=(H2-H3)/(H2 H1)
 390  PRINT "IDEAL COP OF REVERSIBLE HEAT PUMP=";COP
 400  REM FIND ENERGY TO DRIVE HEAT PUMP
 410  EHP=E3/(0.3*COP)
 420  C3=(EHP*0.05)/3600
 430  REM FIND ENERGY TO BE SUPPLIED BY PANELS HENCE AREA NEEDED
 440  ESP=(E3-EHP)
 450  ASP=(5*ESP)/E2
 460  PRINT "HOW MANY YEARS TO ALLOW FOR CAPITAL COSTS?"
 470  INPUT Y
 480  S12=(C1-C2)*Y
 490  S13=(C1-C3)*Y
 500  PRINT
 510  PRINT "ALL ELECTRIC WATER HEATING COST=";C1;"POUNDS/ANNUM"
```

```
520 PRINT "SOLAR PANEL WATER HEATING COST=";C2;"POUNDS/ANNUM"
530 PRINT "HEAT PUMP WITH PANELS HEATING COST=";C3;"POUNDS/ANNUM"
540 PRINT
550 PRINT "PANEL AREA NEEDED WITH HEAT PUMP=";ASP;"M↑2"
560 PRINT
570 PRINT "CAPITAL SAVED IN";Y;"YEARS WITH SOLAR PANELS=";S12;"POUNDS"
580 PRINT "CAPITAL SAVED IN";Y;"YEARS WITH HEAT PUMP=";S13;"POUNDS"
590 END
```

```
RUN
EXAMPLE 6.6

WHAT IS THE DAILY FLOW RATE IN LITRES?
? 250
WHAT IS THE PANEL AREA IN M↑2?
? 5
FIND COP OF HEAT PUMP
 0.6921
ENTROPY?
? 0.6921
IDEAL COP OF REVERSIBLE HEAT PUMP= 6.676927
HOW MANY YEARS TO ALLOW FOR CAPITAL COSTS?
? 10

ALL ELECTRIC WATER HEATING COST= 185.414931POUNDS/ANNUM
SOLAR PANEL WATER HEATING COST= 141.614931POUNDS/ANNUM
HEAT PUMP WITH PANELS HEATING COST= 92.5650021POUNDS/ANNUM

PANEL AREA NEEDED WITH HEAT PUMP= 10.5993069M↑2

CAPITAL SAVED IN 10YEARS WITH SOLAR PANELS= 438POUNDS
CAPITAL SAVED IN 10YEARS WITH HEAT PUMP= 928.499284POUNDS

STOP AT 590
```

Program nomenclature

E2	energy from panel
COP	coefficient of performance
EHP	heat pump energy input
C1	running cost
C2	running cost with panel
C3	running cost with panel and heat pump
ESP	panel energy
ASP	panel area
S12	saving with panel
S13	saving with heat pump and panel

Program notes

(1) The total energy needed to heat F litres of water per day in a year is calculated and costed in line 60.

(2) The energy obtained from the solar panel of area A is calculated in line 110. The efficiency of 20 per cent allows for the efficiency of the whole system as an energy transfer method.

(3) The coefficient of performance of the heat pump is found in lines

260 to 380 and is a similar calculation to that of example 4.4 for the refrigerator. The efficiency of 30 per cent allows for the whole heat pump system and the c.o.p. is effectively about 2.
(4) The financial calculation based on electrical water heating shows a more optimistic picture than using gas heating but no inflation allowance has been made. The results clearly show that the payback period for such an installation is likely to be far greater than that found for wall insulation in example 7.1.
(5) By changing panel areas and water flow rates and temperatures other more (or less) favourable results may be obtained.

Example 6.7 Multi-stage reciprocating air compressor

Problem 4.5 will have demonstrated that as the pressure ratio in a reciprocating compressor of fixed clearance volume (as a fraction of

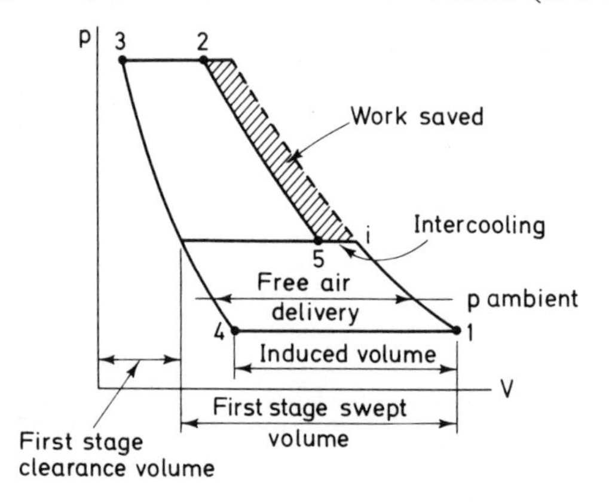

Figure 6.6 Two-stage air compression

swept volume) is increased so the volume of air induced and delivered is reduced. This is expressed by the volumetric efficiency of the compressor

$$\eta_{VOL} = \frac{\text{Volume of air induced, measured at a standard } p \text{ and } T}{\text{swept volume}} \qquad (6.5)$$

When this efficiency becomes unacceptably low due to high pressure ratio it is necessary to perform the compression in a number of stages. Between stages the air is cooled to reduce its volume and so save work in subsequent stages. Perfect inter-cooling reduces the temperature of the air to the inlet temperature ($T_5 = T_1$ in Figure 6.6). Volumetric efficiency of a multi-stage compressor is measured with the first stage dimensions since the air induced will pass to subsequent stages without problems. The volume of air induced is usually measured at free air conditions of approximately 101 kN/m^2, 15°C and is called the free air delivery. This volume is not equal to $(V_1 - V_4)$ on Figure 6.6 since $p_1 < 101$ kN/m^2 and $T_1 > 15$°C.

It is found by differentiation, with perfect inter-cooling and reversible, polytropic compression and expansion with the same index n, that compressor power is minimised if the pressure ratio for each stage is equal.

In Figure 6.6

$$\frac{p_i}{p_1} = \frac{p_2}{p_i} \quad \text{or} \quad r_{p_{\text{stage}}} = \sqrt{r_{p_{\text{overall}}}}$$

and for N stages

$$r_{p_{\text{stage}}} = \sqrt[N]{r_{p_{\text{overall}}}} \tag{6.6}$$

This results in equal work for each stage and thus the total power

$$\dot{W}_x = N\,\dot{W}_{x_{\text{stage}}} = N\,\dot{m}\left[\frac{n}{n-1}\right]RT_1\left[1 - (r_{p_{\text{stage}}})^{\frac{n-1}{n}}\right]$$

(from equation 4.13)

where $\dot{m}$ is the air mass flow rate.

Write a program to verify that the specific work is minimised in a two-stage reciprocating air compressor (with perfect inter-cooling and reversible polytropic expansion and compression) when equal pressure ratios are chosen. Evaluate the program with compression from 100 kN/m^2, 25°C to 6400 kN/m^2 with $n = 1.3$ and $R = 0.287$ kJ/kg K.

The necessary reading for this example is in section 4.11.

```
LIST
 10  PRINT "EXAMPLE 6.7"
 20  PRINT
 30  P1=100
 40  P2=6400
 50  R=0.287
 60  T1=25+273
 70  N=1.3
 80  I=(N-1)/N
 90  DIM PI[8],R1[8],R2[8],W[8]
 100  PRINT "1ST STAGE RP      2ND STAGE RP      WORK(KJ/KG)"
 110  PRINT
 120  FOR J=0 TO 7
 130   READ PI[J]
 140   R1[J]=PI[J]/P1
 150   R2[J]=P2/PI[J]
 160   W1=(N*R*T1*(1-(R1[J]↑I)))/(N-1)
 170   W2=(N*R*T1*(1-(R2[J]↑I)))/(N-1)
 180   W[J]=W1+W2
 190   PRINT R1[J],R2[J],W[J]
 200  NEXT J
```

```
210  A=0.5
220  IP=(6400/100)^A
230  PRINT
240  PRINT "IDEAL STAGE PRESSURE RATIO=";IP
250  PRINT "DOES THE TABULATED PRESSURE RATIO FOR MINIMUM WORK"
255  PRINT "AGREE WITH THE IDEAL VALUE?"
260  DATA 200,300,400,600,800,1000,2000,3000
270  END
```

```
RUN
EXAMPLE 6.7


1ST STAGE RP      2ND STAGE RP      WORK(KJ/KG)

 2                32                -518.310491
 3                21.3333333        -487.307711
 4                16                -471.850883
 6                10.6666667        -459.135969
 8                8                 -456.495588
 10               6.4               -458.083937
 20               3.2               -483.371648
 30               2.13333333        -512.645419


IDEAL STAGE PRESSURE RATIO= 8
DOES THE TABULATED PRESSURE RATIO FOR MINIMUM WORK
AGREE WITH THE IDEAL VALUE?


STOP AT 270
```

Program nomenclature

P1 inlet pressure
P2 delivery pressure
R specific gas constant
T1 inlet temperature
N index of compression and re-expansion
PI interstage pressure
R1 first stage pressure ratio
R2 second stage pressure ratio
W1 first stage work transfer
W2 second stage work transfer
W total work transfer
IP ideal interstage pressure

Program notes

(1) The values of intermediate pressure to be tried are given in line 260 and the work transfer in each stage is evaluated in lines 160 and 170.
(2) The results clearly show agreement with the ideal value.
(3) The program could be extended to determine the work saved by inter-cooling, and the interstage pressure for minimum work with inter-cooling that is not ideal, i.e. $T_5 > T_1$.

Example 6.8 Air conditioning plant and load

In air conditioning plant design (Figure 6.7) the load (due to solar warming of the building, infiltration of outside air, cooking, dancing, working, etc.) is calculated in two parts, that which raises the humidity

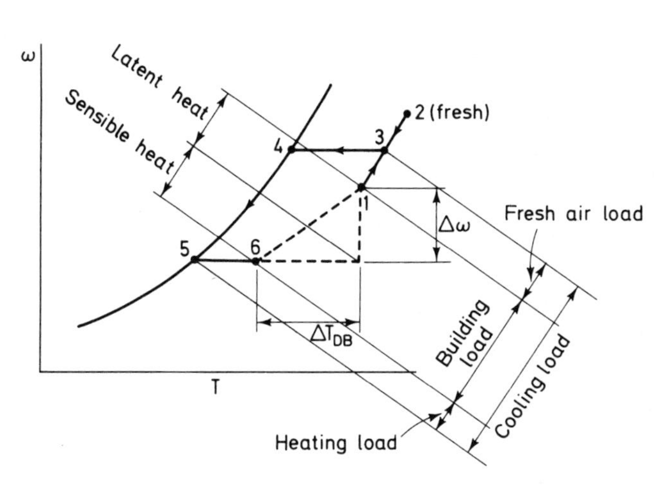

Figure 6.7 Air conditioning loads

(latent heating) and that which raises the dry bulb temperature (sensible heating). This information is used to determine the slope of the process path 6–1 (called the ratio line). The air is delivered from the conditioner at state 6 and removed from the space at state 1 (and then mixed with the necessary amount of fresh air before being recycled through the air conditioner). The temperature change associated with the process 6–1 is a measure of how closely the internal conditions are controlled. The allowed change in dry bulb temperature may be between 2°C and 10°C. If the change is small, the flow rate of conditioned air is large, requiring either high velocity or large ducts. The critical points in the design will be permissible velocity, pressure loss in ducting and space available for ducting.

Write a program, using the data below, to determine circular duct size and the pressure drop per metre length of duct in N/m^2 with a velocity of 8 m/s and a dry bulb temperature rise of 6°C (use the program of example 3.7 to determine the psychrometric properties).

Data

Fresh air state (2): 30°C dry bulb, $\phi = 80\%$
Recirculated air state (1): 25°C dry bulb, $\phi = 71\%$

Ratio $\dfrac{\text{humidity change}}{\text{dry bulb } T \text{ change}} = \dfrac{\Delta\omega}{\Delta T_{DB}} = 0.0008/°\text{C}$

Delivered air state (6): 19°C dry bulb, $\phi = 66\%$

Load: 100 kW

Mass of fresh air/mass of recirculated air = 0.25

Darcy equation for head loss/unit length: $h_f = \dfrac{4fV^2}{2gd}$; $f = 0.008$

The necessary reading for this example is in section 4.13.

```
LIST
 10  PRINT "EXAMPLE 6.8"
 20  PRINT
 30  PRINT "FIND PROPERTIES OF RECIRCULATED AIR(NO DEW POINT)"
 40  GOSUB 4030
 50  H1=H
 60  W1=W
 70  PRINT "FIND PROPERTIES OF FRESH AIR(NO DEW POINT)"
 80  GOSUB 4070
 90  H2=H
 100  W2=W
 110  H3=(4*H1+H2)/5
 120  W3=(4*W1+W2)/5
 130  PRINT "FIND PROPERTIES OF DELIVERED AIR(WITH DEW POINT)"
 140  GOSUB 4070
 150  H6=H
 160  V6=V
 170  T5=C
 180  PRINT "FIND PROPERTIES AT STATE 5 WITH TDB=T5,PHI=1"
 190  GOSUB 4070
 200  H5=H
 210  W5=W
 220  REM FIND AIR FLOW RATE,DUCT SIZE,PRESSURE LOSS /METRE
 230  F=V6*100/(H1-H6)
 240  PRINT "AIR FLOW RATE FROM CONDITIONER =";F;"M↑3/S"
 250  VEL=8
 260  A=F/VEL
 270  D=((4*A)/3.142)↑0.5
 280  PD=(4*0.008*(VEL↑2))/(2*V6*D)
 290  PRINT
 300  PRINT "DUCT DIAMETER=";D;"M"
 310  PRINT
 320  PRINT "PRESSURE DROP=";PD;"N/M↑2/M"
 330  PRINT
 340  REM FIND COOLING POWER WITH COP REFRIGERATOR OF 3 AND HEATING POWER
 350  QC=(((H5-H3)+(W3-W5)*4.2*T5)*(F/V6))/3
 360  QH=(H6-H5)*(F/V6)
 370  PRINT "COOLING POWER=";QC;"KW"
 380  PRINT
 390  PRINT "HEATING POWER=";QH;"KW"
 400  END
 4030  DIM T[8],P[8]
 4040  FOR N=0 TO 7
 4050   READ T[N],P[N]
 4060  NEXT N
 4070  REM INTERPOLATE FOR PG AT TDB
```

FOR THE PSYCHROMETRIC PROPERTIES PROGRAM SEE EXAMPLE 3.7

```
4420LIST
 4420  C=C+Z*T[J]
 4430 NEXT J
 4440  PRINT "DEW POINT TEMPERATURE=";C;" C"
```

```
4450  RETURN
4460  DATA Ø.Ø1,Ø.611
4470  DATA 5,Ø.872
4480  DATA 1Ø,1.227
4490  DATA 15,1.7Ø4
4500  DATA 2Ø,2.337
4510  DATA 25,3.166
4520  DATA 3Ø,4.241
4530  DATA 35,5.622


RUN
EXAMPLE 6.8

FIND PROPERTIES OF RECIRCULATED AIR(NO DEW POINT)
TDB?
? 25
PG= 3.166KN/M↑2
PHI?
? .71
PS= 2.24786KN/M↑2
ENTHALPY DATUM IS WATER AT Ø.Ø1 C
SPECIFIC ENTHALPY,H= 61.2Ø85897KJ/KG DRY AIR
SPECIFIC VOLUME,V= Ø.863226371M↑3/KG DRY AIR
SPECIFIC HUMIDITY,W= Ø.Ø141119225
IF PS<Ø.611 DEW POINT TEMP IS BELOW O C AND PROGRAM INVALID
DO YOU WANT DEW POINT TEMPERATURE?YES,TYPE 1,NO,TYPE 2
? 2
FIND PROPERTIES OF FRESH AIR(NO DEW POINT)
TDB?
? 3Ø
PG= 4.241KN/M↑2
PHI?
? .8
PS= 3.3928KN/M↑2
ENTHALPY DATUM IS WATER AT Ø.Ø1 C
SPECIFIC ENTHALPY,H= 85.4Ø89Ø66KJ/KG DRY AIR
SPECIFIC VOLUME,V= Ø.887971474M↑3/KG DRY AIR
SPECIFIC HUMIDITY,W= Ø.Ø215488Ø21
IF PS<Ø.611 DEW POINT TEMP IS BELOW O C AND PROGRAM INVALID
DO YOU WANT DEW POINT TEMPERATURE?YES,TYPE 1,NO,TYPE 2
? 2
FIND PROPERTIES OF DELIVERED AIR(WITH DEW POINT)
TDB?
? 19
PG= 2.19612Ø27KN/M↑2
PHI?
? .66
PS= 1.44943938KN/M↑2
ENTHALPY DATUM IS WATER AT Ø.Ø1 C
SPECIFIC ENTHALPY,H= 42.Ø9Ø3124KJ/KG DRY AIR
SPECIFIC VOLUME,V= Ø.839Ø84151M↑3/KG DRY AIR
SPECIFIC HUMIDITY,W= Ø.ØØ9Ø2674575
IF PS<Ø.611 DEW POINT TEMP IS BELOW O C AND PROGRAM INVALID
DO YOU WANT DEW POINT TEMPERATURE?YES,TYPE 1,NO,TYPE 2
? 1
INTERPOLATE FOR DEW POINT AT PS
DEW POINT TEMPERATURE= 12.5Ø9ØØ57 C
FIND PROPERTIES AT STATE 5 WITH TDB=T5,PHI=1
TDB?
? 12.51
PG= 1.449176KN/M↑2
PHI?
? 1
PS= 1.449176KN/M↑2
ENTHALPY DATUM IS WATER AT Ø.Ø1 C
SPECIFIC ENTHALPY,H= 35.4222466KJ/KG DRY AIR
SPECIFIC VOLUME,V= Ø.82Ø432481M↑3/KG DRY AIR
SPECIFIC HUMIDITY,W= Ø.ØØ9Ø25Ø8218
IF PS<Ø.611 DEW POINT TEMP IS BELOW O C AND PROGRAM INVALID
DO YOU WANT DEW POINT TEMPERATURE?YES,TYPE 1,NO,TYPE 2
? 2
```

```
AIR FLOW RATE FROM CONDITIONER = 4.38891087M↑3/S

DUCT DIAMETER= 0.835719031M

PRESSURE DROP= 1.46027331N/M↑2/M

COOLING POWER= -52.79592KW

HEATING POWER= 34.8779633KW

STOP AT 400
```

Program nomenclature

H	specific enthalpy
W	specific humidity
V	specific volume
PHI	relative humidity
TDB	dry bulb temperature
F	air flow rate
A	duct area
D	duct diameter
PD	duct pressure drop
COP	coefficient of performance
QC	cooling power
QH	heating power

Program notes

(1) The program from example 3.7 for psychrometric properties has been used as a subroutine and is called four times, in lines 40, 80, 140 and 190. The program has been modified in two places. The first two title lines have been omitted (4010 and 4020) and line 4450 has been changed from END to RETURN which allows the main program to proceed from the point where it was instructed to GOSUB. In lines 80, 140 and 190 the subroutine is entered at line 4070 rather than 4030 since the steam data of the subroutine has previously been read and stored in lines 4030 to 4060, called in line 40.
(2) Lines 110 and 120 are mixing calculations to determine the input state to the conditioner.
(3) The air flow rate is found by dividing the building load by the specific enthalpy change allowed by the choice of delivered air state which was 6°C below the recirculation state with a velocity of 8 m/s.
(4) The resulting duct size will need to be fitted into the building and if too large the velocity would need to be increased. Unfortunately this will result in a higher pressure drop which is already above the maximum normally permitted (1 N/m^2 per metre). Thus the flow rate

should be reduced by increasing the temperature differential between delivered air and recirculated air, which means a less closely controlled environment. These needs may be conflicting. The program could be run several times to find a compromise.

(5) The total power input to the system would be the heating power, the cooling power and the fan power to circulate the air (obtained from flow rate, pressure loss and duct length, other duct fitting losses, etc.)

Example 6.9 Spark ignition engine performance

The results of a trial on a 1250 cc petrol engine with spark-ignition are given below (in British units to give mpg and mph data later). Write a program to obtain torque, power and specific fuel consumption at the various engine speeds (in ft lbf, horsepower and lb/hp hour respectively). Arrange the results in a data table for interpolation and also plot curves of the results (power, torque and sfc as ordinate versus speed).

Trial results

Speed N rev/ min	*Brake load W lbf*	*Time to use 0.1 lb fuel S sec*	*W lbf*	*S sec*	*W lbf*	*S sec*	*W lbf*	*S sec*
	Full Load		*¾ Load*		*½ Load*		*¼ Load*	
5500	53	11.4	39.6	14.7	26.7	19.8	13.8	27.6
5000	57.8	12.1	43.1	16.0	29.4	21.4	15.2	30.3
4500	60.7	13.4	45.5	17.4	30.3	24.7	15.3	34.8
4000	60.4	15.8	46.0	20.2	30.2	29.0	15.9	39.7
3500	62.3	18.5	47.3	23.3	31.5	33.0	15.9	46.5
3000	65.1	21.0	47.3	29.0	31.5	39.6	16.1	55.1
2500	65.1	24.2	48.3	34.8	32.8	46.2	17.0	63.5
2000	62.0	29.3	47.3	44.4	35.0	56.8	16.0	84.3
1500	54.3	41.1	41.0	61.1	28.0	83.3	14.0	100.0
1000	47.3	64.5	34.7	97.4	24.2	122.3	10.5	174.8

Brake arm = 1 ft and torque = brake load × brake arm

$$\text{Horsepower} = \frac{2\pi \times \text{speed in rev/min} \times \text{torque in ft lbf}}{33\,000}$$

$$\text{Specific fuel consumption} = \frac{0.1 \times 3600}{\text{power} \times \text{time to consume 0.1 lb of fuel}}$$

```
LIST
 10  PRINT "EXAMPLE 6.9"
 20  PRINT
 30  DIM N[40],W[40],S[40],T[40],P[40],SFC[40]
 40  PRINT "SPEED(REV/MIN)  TORQUE(FT LBF)  POWER(HP)        SFC(LB/HP H)"
 50  PRINT "FULL LOAD"
```

```
60  FOR J=0 TO 9
70   GOSUB 2000
80  NEXT J
90  PRINT
100  PRINT "THREEQUARTER LOAD"
110  FOR J=10 TO 19
120   GOSUB 2000
130  NEXT J
140  PRINT
150  PRINT "HALF LOAD"
160  FOR J=20 TO 29
170   GOSUB 2000
180  NEXT J
190  PRINT
200  PRINT "QUARTER LOAD"
210  FOR J=30 TO 39
220   GOSUB 2000
230  NEXT J
240  PRINT
250  END
2000  READ N[J],W[J],S[J]
2010  T[J]=1*W[J]
2020  P[J]=(2*3.142*N[J]*T[J])/33000
2030  SFC[J]=(0.1*3600)/(S[J]*P[J])
2040  PRINT N[J],T[J],P[J],SFC[J]
2050  RETURN
3000  DATA 5500,53,11.4
3010  DATA 5000,57.8,12.1
3020  DATA 4500,60.7,13.4
3030  DATA 4000,60.4,15.8

       ETC

3360  DATA 2500,17,63.5
3370  DATA 2000,16,84.3
3380  DATA 1500,14,100
3390  DATA 1000,10.5,174.8
```

```
RUN
EXAMPLE 6.9

SPEED(REV/MIN)   TORQUE(FT LBF)   POWER(HP)        SFC(LB/HP H)
FULL LOAD
 5500             53               55.5086667       0.5689012
 5000             57.8             55.0326061       0.540626153
 4500             60.7             52.0143818       0.516504680
 4000             60.4             46.0064970       0.495252011
 3500             62.3             41.5220061       0.468654126
 3000             65.1             37.1898545       0.4609552
 2500             65.1             30.9915455       0.480002944
 2000             62               23.6126061       0.520344488
 1500             54.3             15.5100545       0.564738445
 1000             47.3             9.00706667       0.619668484

THREEQUARTER LOAD
 5500             39.6             41.4744          0.590479812
 5000             43.1             41.0364242       0.548293386
 4500             45.5             38.9893636       0.5306487
 4000             46               35.03806         0.508640657
 3500             47.3             31.5247333       0.490111799
 3000             47.3             27.0212          0.459409393
 2500             48.3             22.9937273       0.449897812
 2000             47.3             18.0141333       0.450097041
 1500             41               11.71109         0.503111145
 1000             34.7             6.60772121       0.559360549

HALF LOAD
 5500             26.7             27.9638          0.650191254
 5000             29.4             27.9923636       0.600964968
 4500             30.3             25.9643455       0.561342816
```

```
 4000          30.2          23.0032485       0.539653915
 3500          31.5          20.9942727       0.519622235
 3000          31.5          17.99509         0.505188284
 2500          32.8          15.6147879       0.499027451
 2000          33            12.568           0.504298868
 1500          28            7.99781818       0.540363458
 1000          24.2          4.60826667       0.638761072

QUARTER LOAD
 5500          13.8          14.4532          0.902463
 5000          15.2          14.4722424       0.820963868
 4500          15.3          13.1107091       0.789036467
 4000          15.9          12.1109818       0.748742770
 3500          15.9          10.5971091       0.730570519
 3000          16.1          9.19749091       0.710364966
 2500          17            8.09303030       0.700515274
 2000          16            6.09357576       0.7008139
 1500          14            3.998909         0.900245522
 1000          10.5          1.99945455       1.03002920

STOP AT 250
```

Program nomenclature

N engine speed
W brake load
S time to consume fuel
T torque
P power
SFC specific fuel consumption

Program notes

(1) The program is straightforward. The data is read in and then processed four times by the subroutine in lines 2000 to 2050.
(2) The results should be plotted to enable the engine performance to be visualised.
(3) Problem 6.10 should now be tried to see how the engine data may be matched to a motor car.

PROBLEMS

(6.1) In order to overcome the problems of wet steam at the end of expansion in high pressure steam plant cycles (see problem 4.4) reheating is used part way through the expansion. Reheating also improves cycle efficiency. The expansion is split into two parts (Figure 6.8) and reheating can be used to the metallurgical limit of the material in the reheater. Write a program to determine the efficiency of the reheat cycle using a maximum cycle and reheat temperature of 600°C and a cycle maximum pressure of 5 MN/m^2. Find the optimum

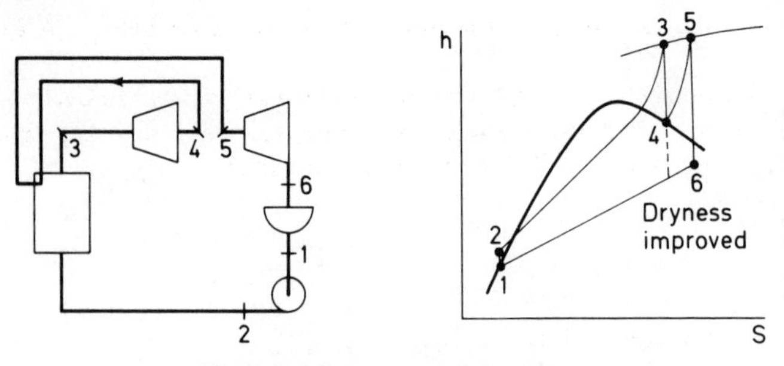

Figure 6.8 Steam cycle with reheat

pressure for reheating (maximum efficiency) and compare the dryness fraction at the end of expansion with that in problem 4.4. Expansions should be reversible and adiabatic. The condenser pressure is 0.004 MN/m^2.

(6.2) Write a program to determine the optimum siting of two feed heaters in a Rankine cycle (see example 6.1). Use the data of example 6.1 and compare the results obtained with those of that example. Figure 6.9 shows the plant. The same assumptions may be used about the bleed steam mass flow rates as in example 6.1.

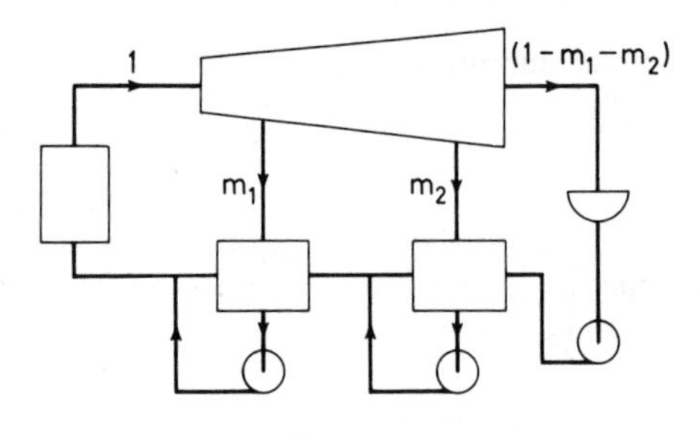

Figure 6.9

(6.3) In a trial on a small gas turbine plant consisting of a compressor, a combustion chamber and a turbine driving the compressor, a brake is used to absorb the net power and the air mass flow rate is measured by means of an inlet nozzle for which

$$\dot{m} = 0.0748 \sqrt{\frac{P}{T}} \sqrt{h} \text{ kg/s}$$

where p = air inlet pressure in kN/m^2
T = ambient temperature in K
h = air meter depression in mm oil

The results of the trial and other data are given below. Write a program:

(a) to determine the power output, overall efficiency, air-fuel ratio, turbine entry temperature, compressor pressure ratio, turbine pressure ratio and the brake specific fuel consumption; and

(b) to enable energy balances to be drawn up.

Data

s.g. of mercury 13.6
s.g. of oil in meters 0.85
s.g. of fuel 0.848
Calorific value of fuel 46 000 kJ/kg
Turbine mechanical efficiency 0.98
Compressor mechanical efficiency 0.97
Brake equation:

$$\text{Power in kW} = \frac{\text{Load (N)} \times \text{Speed (rev/min)}}{26\,850}$$

For air; $\gamma = 1.4$ and $C_p = 1.01$ kJ/kg K
For gas; $\gamma = 1.33$ and $C_p = 1.15$ kJ/kg K

Trial results

Brake load (N)	50	100	150	200	250	300
Brake speed (rev/min)	3040	3025	3025	3025	3020	3010
Air inlet temperature (°C)	9	9.5	10	10	10.5	11
Ambient pressure (mm Hg)	775	775	775	775	775	775
Inlet nozzle depression (mm oil)	271	260	255	248	240	233
Compressor delivery gauge pressure (kN/m²)	188	188	190	190	193	194
Compressor delivery temperature (°C)	152	151	151	151	151	151
Pressure loss in combustion chamber (mm Hg)	145	140	135	128	126	117
Turbine exit temperature (°C)	390	412	430	455	475	502
Turbine exit gauge pressure (mm oil)	90	75	70	58	57	57
Time to use 2 dm³ of fuel (seconds)	218	208	196	184	171	160

(6.4) A gas turbine plant for a pumping station consists of a compressor, a heat exchanger, a combustion chamber and a turbine which drives the compressor and the pump. Write a program to determine the efficiency of the plant (given by pump power/rate of energy input in fuel). All components are adiabatic. Evaluate the program using the data below. Figure 6.10 shows the plant.

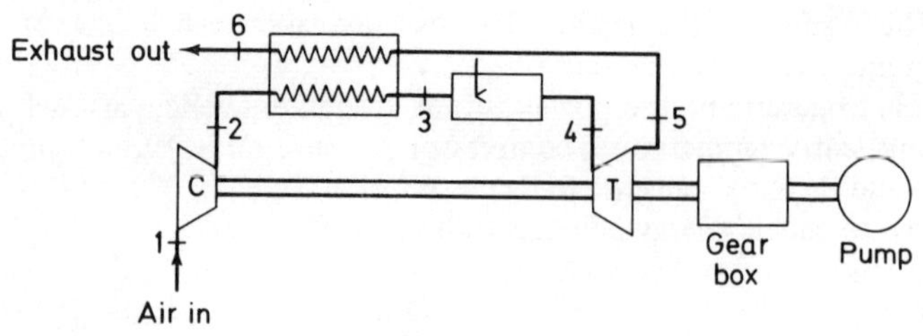

Figure 6.10

Data

Compressor entry state	100 kN/m^2, 290 K
Compressor efficiency	0.84
Compressor pressure ratio	5
Heat exchanger efficiency	0.68
Pump power requirement	2300 kW
Turbine outlet temperature	680 K
Air mass flow rate	30 kg/s
Calorific value of fuel	44 000 kJ/kg
Combustion efficiency	96%
Pressure losses in heat exchanger	5 kN/m^2
Pressure loss in combustion chamber	15 kN/m^2

For air; $\gamma = 1.4$, $C_p = 1.01$ kJ/kg K
For gas; $\gamma = 1.33$, $C_p = 1.15$ kJ/kg K

(6.5) An aircraft jet propulsion engine consists of an intake diffuser, a compressor, a combustion chamber, a turbine driving the compressor and a convergent-divergent nozzle expanding the exhaust gases to the ambient pressure. Write a program to determine the thrust of the engine and the efficiency (given by thrust $\times$ velocity/rate of energy input in fuel). All components are adiabatic. Evaluate the program using the data below. Figure 6.11 shows the engine.

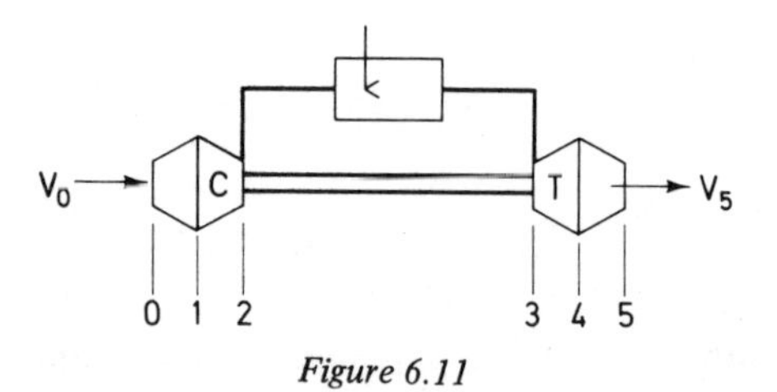

Figure 6.11

Data

Aircraft speed	190 m/s
Ambient state	60 kN/m^2, 260 K
Intake efficiency	0.84

Compressor efficiency	0.82
Compressor delivery stagnation pressure	480 kN/m^2
Turbine efficiency	0.85
Turbine exit stagnation temperature	900 K
Combustion chamber pressure loss	10 kN/m^2
Combustion efficiency	0.96
Calorific value of fuel	45 MJ/kg
Air mass flow rate	37 kg/s
Nozzle efficiency	0.9

γ and C_p values as in problem 6.4

(6.6) Modify the program of example 6.4 to allow for heat transfer through the cylinder walls, head and piston top. Assume that the mean surface temperature is 380 K and that the cylinder head and piston top are flat. The surface heat transfer coefficient h_θ (for that particular engine at the test speed) is given by

$$h_\theta = 0.45\sqrt{p_\theta T_\theta} \text{ W/m}^2\text{K with } p_\theta \text{ in kN/m}^2,\ T_\theta \text{ in K.}$$

The heat transfer area = piston top + cylinder head + (exposed wall area)$_\theta$

$$A_\theta = \frac{2\pi}{4}(\text{bore})^2 + \pi\,(\text{bore}) \times (\text{exposed wall height})_\theta$$

Thus $$\dot{Q}_\theta = h_\theta\, A_\theta\, (T_\theta - T_{\text{wall}}) \text{ watts}$$

Heat transfer per degree of crank angle will therefore be

$$Q_\theta = \dot{Q}_\theta \times \text{time per degree of crank angle (joules)}$$

This method includes an allowance for radiation.

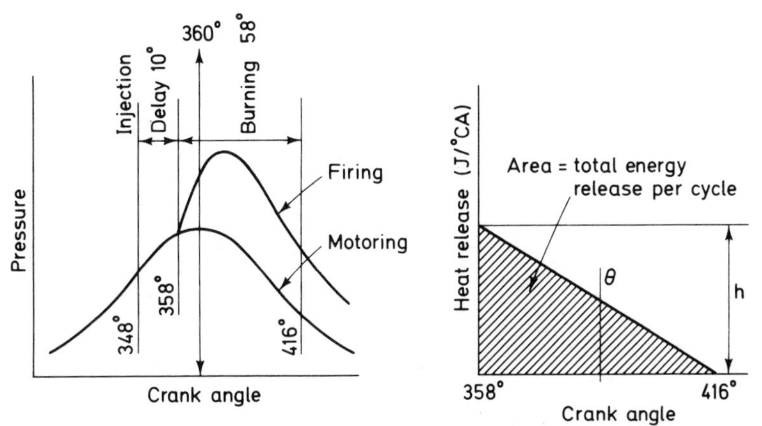

Figure 6.12 Combustion in compression ignition engines

(6.7) Modify the program of problem 6.6 to allow for heat release as the fuel is burned. Figure 6.12 shows the point of injection after which there is a delay period before the fuel commences to burn. After this delay there is a rapid energy release when the fuel already injected ignites, followed by a slowing of the energy release rate as the fuel continues to be injected and burnt. After injection ceases, burning continues till the fuel is all consumed but the energy release rate continues to fall as the fuel is consumed. There are several ways of representing this energy release rate and in this problem a triangular energy release diagram is suggested.

The fuel flow rate is 0.02 g/cycle with calorific value 42 000 J/g giving an energy release of 840 J/cycle. Thus for the triangular energy release diagram

$$h = \frac{840}{\frac{1}{2}(416-358)} \text{ J/degree crank angle}$$

Thus at any crank angle θ the energy input per degree of crank angle is

$$\left[\frac{416-\theta}{416-358}\right] \times \frac{840}{\frac{1}{2}(416-358)} = 0.5(416-\theta)\ \text{J/°CA}$$

So energy input for each increment of crank angle

$$= [0.5(416-\theta) \times \text{increment size in degrees}]\ \text{J}$$

If a need is felt to make the program more realistic and more complex the mass in the cylinder may be increased uniformly over the injection period from a crank angle of 348° for 30 degrees and the value of C_v may be changed to 0.87 kJ/kg K from a crank angle of 358°.

(6.8) An alternative refrigeration plant layout using a flash chamber is shown in Figure 6.13. Examine the plant and decide (with the same condenser and evaporator conditions as in example 6.5) whether the

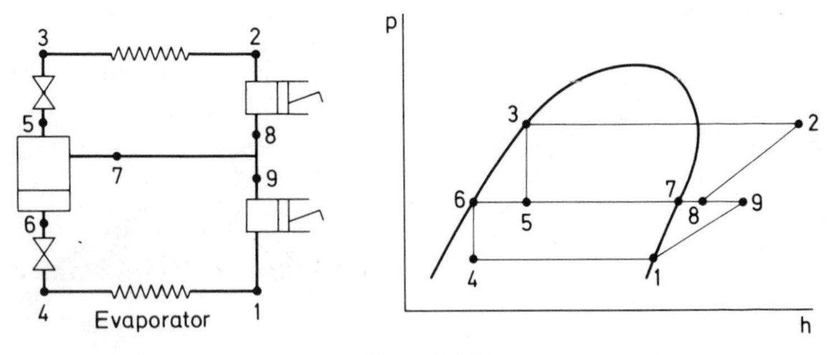

Figure 6.13

coefficient of performance should be better or worse than that of the layout considered in example 6.5. Write a program to verify your decision.

(6.9) Write a program to determine the minimum power needed by a reciprocating compressor for a free air delivery of 0.5 m^3/s measured at 101 kN/m^2, 15°C, which is compressed from 100 kN/m^2, 25°C to 20 MN/m^2. Choose 2, 3 or 4 stages with the clearance volume of the first stage 5 per cent of the swept volume. The program should also determine the volumetric efficiency based on the free air delivery for each case. For all compression and expansion processes (which may be considered reversible and polytropic) $n = 1.3$, and for air $R = 0.287$ kJ/kg K.

(6.10) The engine of example 6.9 is fitted into a motor car. Continue the program to determine the fuel consumption in mpg at a steady speed of V mph on level ground in top gear. Evaluate the program at 56 mph, for which test results suggest 43.5 mpg.

Data

At steady speed, Resistance = Rolling + Drag

Rolling resistance = $(25 + 2.5 \times 10^{-6}\ V^{3.7})$ lbf with V in mph

Drag = $\frac{1}{2} C_D A \rho V^2$

Drag coefficient, $C_D = 0.45$

Frontal area, $A = 19$ ft^2

Air density, $\rho = 0.076$ lb/ft^3

1 gallon = 7.9 lb of fuel

Vehicle mass 2040 lb

Wheel diameter 22 in

Gravitational acceleration = 32.2 ft/s^2

Force/mass relation 1 lbf = $32.2 \dfrac{\text{lb ft}}{\text{s}^2}$

Transmission efficiency = 96%

Gear ratios:	
1st	3.76
2nd	2.213
3rd	1.404
4th	1
Axle ratio	4.111

1 horsepower = 550 ft lbf/s

(6.11) Rework example 6.8 for a range of options:

(a) vary the dry bulb temperature change allowed in the room from 2°C to 12°C in 2°C steps with a fixed velocity of 8 m/s and determine the duct size and pressure loss per metre;

(b) vary the dry bulb temperature change allowed in the room from 2°C to 10°C in 2°C steps, with the duct size held constant at the value found in example 6.8, and determine the velocity and pressure loss per metre.

Data for the ratio line with changing temperature rise is as follows:

T°*C*	*Dry bulb temperature* (°*C*)	*Relative humidity* %
2	23	70
4	21	70
6	19	66
8	17	64
10	15	58

(6.12) A cooling tower is used to provide cooling water in situations where a restricted water supply is available. After being used as a coolant the warmed water is passed down the cooling tower (Figure 6.14) inside which it meets a counter flow of atmospheric air (forced draught by fan or natural draught). Part of the water evaporates and the air becomes hotter and more humid. The evaporation of some water

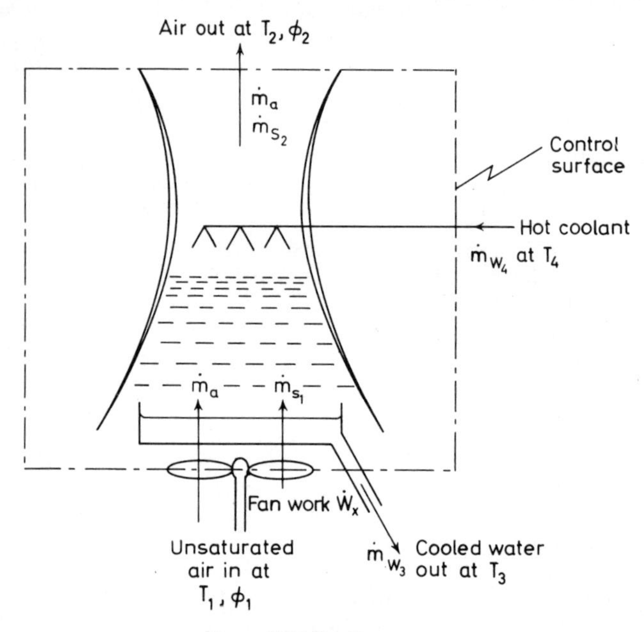

Figure 6.14 Cooling tower

takes energy from the remaining water which is therefore cooled and available for reuse as a coolant. The evaporated water has to be supplied as 'make-up'.

By application of the steady flow energy equation to the adiabatic cooling tower

$$-\dot{W}_x = \dot{m}_a (h_2 - h_1) + \dot{m}_{w_3} h_{w_3} - \dot{m}_{w_4} h_{w_4}$$

and by the continuity equation

$$\dot{m}_a(\omega_2 - \omega_1) = \dot{m}_{w4} - \dot{m}_{w3}$$

where h_1, h_2, ω_1 and ω_2 are obtained from psychrometric data,

$$h_{w_3} = C_p\, T_3 \text{ and } h_{w_4} = C_p\, T_4 \text{ and } \dot{W}_x \text{ is negative}$$

Write a program to determine the volumetric air flow rate at inlet to the tower and make-up water mass flow rate for a cooling tower to cool 25 kg/s of water from 37°C to 15°C assuming natural draught ($\dot{W}_x = 0$). The air leaves the tower 4°C below the entry temperature of the water and saturated ($\phi = 1$).

Vary the ambient atmospheric conditions to allow for inlet conditions of $T_1 = 5$, 10, 15 and 25°C with $\phi = 0.5$, 0.75 and 1, which could represent an annual variation for some locations.

The following 'examples' are not precisely presented but represent ideas for other problems which would be suitable for simple investigations.

(6.13) Analysis of laboratory tests and energy balances.
(6.14) Steam plant condenser performance. Air leaks have to be counteracted by an extraction pump which removes both air and steam. Consider the rate of loss of steam at various condensing temperatures and the extra fuel needed to heat the make-up water.
(6.15) The exhaust gases from a gas turbine (see problem 6.3) contain considerable potential for use for heating or power generation using an exhaust gas heated steam generator. Diesel engine exhaust gases have similar possibilities.
(6.16) Power available from the tides will depend on the height of the tide, and the time that the power is available will depend on the time of the tides. Consider how to integrate such a scheme with a generating system (diesel engine powered) to give some predetermined power output pattern over a twenty-four hour period for one tidal cycle of twenty-eight days (tide tables required).
(6.17) Binary fluid power cycles (e.g. mercury and steam). Analysis of performance. This will need a further data block.
(6.18) Liquefaction of gases, i.e. low temperature refrigeration problems.
(6.19) Uses of gases for refrigeration applications (e.g. reversed gas turbine plant cycle).
(6.20) Freezing considerations for vegetable, fish and meat storage – freezing times, etc.

Chapter 7

Heat transfer

ESSENTIAL THEORY

7.1 Introduction

There are three different modes of heat transfer: ***conduction, convection*** and ***radiation.*** In many problems all three modes occur; the energy conducted to the surface of a solid which is in contact with a cooler fluid, such as air, will be convected to the air and radiated to the air and any other cooler surroundings. In this chapter each mode is considered separately and in some common combinations.

7.2 Conduction

The fundamental equation for conduction is Fourier's law

$$\dot{Q}'' = -k \frac{\partial t}{\partial n} \tag{7.1}$$

where $\dot{Q}''$ is the heat transfer rate per unit area* in the n direction, k is the thermal conductivity (W/m K), which for isotropic materials is a function of temperature (usually considered constant over limited temperature ranges), and $\partial t/\partial n$ is the temperature gradient in the n direction (which is negative so that the negative sign in the equation gives positive heat transfer with the temperature gradient).

One-dimensional conduction

Fourier's law becomes (for the x direction)

$$\dot{Q}'' = -k \frac{dT}{dx}$$

*In this chapter $\dot{Q}$ is used for heat transfer rate, $\dot{Q}'$ for heat transfer rate per unit length, $\dot{Q}''$ for heat transfer rate per unit area and $\dot{Q}'''$ for heat transfer rate per unit volume.

This can be integrated for the common cases of plane and cylindrical (tubular) walls to give

Single-layer plane wall	$\dot{Q}'' = k\dfrac{\Delta T}{\Delta x} = \dfrac{\Delta T}{(\Delta x/k)}$
Multi-layer plane wall	$\dot{Q}'' = \Delta T / \sum \left(\dfrac{\Delta x}{k}\right)_{\text{layer}}$
Single-layer cylindrical wall	$\dot{Q}' = \dfrac{2\pi k \Delta T}{\ln\left(\dfrac{r_{\text{outer}}}{r_{\text{inner}}}\right)} = \Delta T \Big/ \dfrac{\ln\left(\dfrac{r_{\text{outer}}}{r_{\text{inner}}}\right)}{2\pi k}$
Multi-layer cylindrical wall	$\dot{Q}' = \Delta T \Big/ \sum \left\{\dfrac{\ln\left(\dfrac{r_{\text{outer}}}{r_{\text{inner}}}\right)}{2\pi k}\right\}_{\text{layer}}$

$\dot{Q}'$ = heat transfer rate per unit *length* of tube
ΔT = overall temperature difference. Δx = layer thickness

7.3 Convection

The fundamental convection law is

$$\dot{Q}'' = h\theta \tag{7.2}$$

where θ is the temperature difference between surface and fluid and h is the surface heat transfer coefficient (W/m^2 K).

The determination of h, which depends on the fluid properties and fluid flow pattern, is explained in section 7.8.

For a plane surface–fluid interface	$\dot{Q}'' = \dfrac{\theta}{1/h}$
For a cylindrical surface–fluid interface	$\dot{Q}' = \dfrac{\theta}{1/2\pi r h}$

7.4 Electrical analogy

An analogy between the conduction and convection laws may be made with Ohm's law for electric current flow.

For conduction in a single-layer plane wall

$$\dot{Q}'' = \Delta T \Big/ \left(\frac{\Delta x}{k}\right)$$

is compared with $I = V/R$ and it can be seen that, if ΔT is analogous to V and $\dot{Q}''$ is analogous to I, then $(\Delta x/k)$ is analogous to R. The quantity $(\Delta x/k)$ is called the *thermal resistance.* For the four conduction and two convection cases above the thermal resistance is

Single-layer plane wall	$\frac{\Delta x}{k}$
Multi-layer plane wall	$\Sigma \left(\frac{\Delta x}{k}\right)_{\text{layer}}$
Single-layer cylindrical wall	$\ln\left(\frac{r_{\text{outer}}}{r_{\text{inner}}}\right) \Big/ 2\pi k$
Multi-layer cylindrical wall	$\Sigma \left\{ \ln\left(\frac{r_{\text{outer}}}{r_{\text{inner}}}\right) \Big/ 2\pi k \right\}_{\text{layer}}$
Plane surface–fluid interface	$1/h$
Cylindrical surface–fluid interface	$1/2\pi r h$

Thermal resistances in series can be added in the same manner as electrical resistances in series.

7.5 Overall heat transfer coefficient

In problems in which two fluids are separated by a solid wall (Figure 7.1) the electrical analogy may be used to add the resistances in series

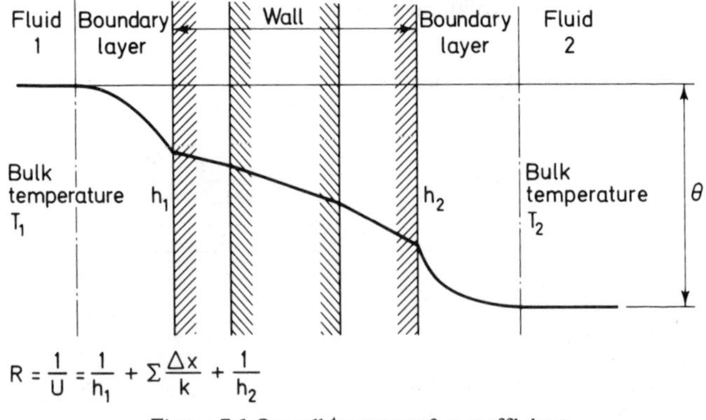

Figure 7.1 Overall heat transfer coefficient

to find the heat transfer rate related to the overall temperature difference θ between the two fluids.

We write

$$\dot{Q} = U A \theta \quad \text{for plane walls} \tag{7.3}$$

and

$$\dot{Q} = U' L \theta \quad \text{for cylindrical walls} \tag{7.4}$$

where U and U' are *overall heat transfer coefficients* given by

$$\frac{1}{U} = \frac{1}{h_1} + \sum \left(\frac{\Delta x}{k}\right)_{\text{layer}} + \frac{1}{h_2} \tag{7.5}$$

and

$$\frac{1}{U'} = \frac{1}{2\pi r_1 h_1} + \sum \left\{ \frac{\ln\left(\frac{r_{\text{outer}}}{r_{\text{inner}}}\right)}{2\pi k} \right\}_{\text{layer}} + \frac{1}{2\pi r_2 h_2} \tag{7.6}$$

Values of U or U' are available for some common constructions such as house walls, double-glazed windows, etc., otherwise they must be calculated.

7.6 General equation for conduction

By considering the thermal equilibrium of a small three-dimensional element of solid, isotropic material it can be shown that

$$\frac{\partial^2 T}{\partial x^2} + \frac{\partial^2 T}{\partial y^2} + \frac{\partial^2 T}{\partial z^2} + \frac{\dot{Q}'''}{k} = \frac{1}{a}\left(\frac{\partial T}{\partial \tau}\right) \tag{7.7}$$

where $\dot{Q}'''$ is the internal heat generation per unit volume (rare except in electric current-carrying conductors), a is the thermal diffusivity of the material, $a = k/\rho C_p$) and $\partial T/\partial \tau$ is the rate of change of temperature with time τ.

7.7 Steady state two-dimensional conduction

In this case the general equation becomes

$$\frac{\partial^2 T}{\partial x^2} + \frac{\partial^2 T}{\partial y^2} + \frac{\dot{Q}'''}{k} = 0$$

If $\dot{Q}''' = 0$ then

$$\frac{\partial^2 T}{\partial x^2} + \frac{\partial^2 T}{\partial y^2} = 0 \tag{7.8}$$

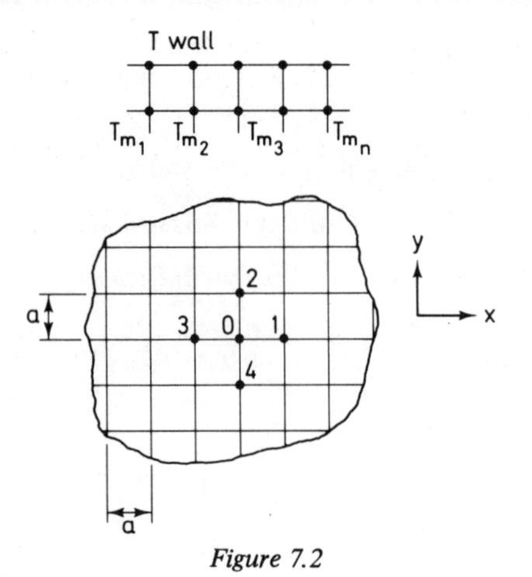

Figure 7.2

The solution to this equation will show the temperature field in the two-dimensional solid and the heat transfer rate may then be estimated from the boundary conditions. A rectangular mesh is drawn on a plan of the surface to be investigated. A portion of such a mesh is shown in Figure 7.2. The temperature at each mesh crossover point is guessed and the correct values are obtained by relaxing the mesh.

Let the temperatures estimated at the points 1, 2, 3 and 4 be T_1, T_2, T_3 and T_4 respectively. The continuously varying temperatures at points 1, 2, 3 and 4 may be represented by a power series, and it can be shown that

$$\frac{\partial^2 T}{\partial x^2} + \frac{\partial^2 T}{\partial y^2} = \frac{T_1 + T_2 + T_3 + T_4 - 4T_0}{a^2}$$

An *approximate* solution to the Laplace equation for two-dimensional temperature distribution is therefore

$$T_1 + T_2 + T_3 + T_4 - 4T_0 = 0 \tag{7.9}$$

To relax the mesh an attempt is made to satisfy this equation at one point in the estimated field. There will probably be a residual at the point, which means the equation will not be equal to zero. This residual is divided by four and the temperature at the point considered is adjusted by the quarter residual (or by such amount as may be judged reasonable, for making an adjustment at one point obviously affects the balance at the surrounding points). This procedure is repeated for each mesh point and then for the whole mesh as many times as is necessary for a satisfactory spread of minimal residuals. The smaller the mesh the more accurate is the result. As the mesh is made smaller the number of mesh points rises rapidly and a computer solution becomes necessary. This can be done in two ways: by *iteration,* which is the method above, or by *matrix inversion.* Iteration is preferable since the same program can be used with small modifications for many problems.

The solution is valid for mesh points within the solid. The heat transfer rate to the solid boundary will be given by

$$\dot{Q} = \Sigma k\,(T_m - T_{\text{wall}}) \text{ per unit height} \tag{7.10}$$

where T_{wall} is the boundary temperature of the solid and T_m is the temperature at each mesh point adjacent to the boundary (Figure 7.2).

In many problems it is necessary to allow for convection at the boundary for which the surface heat transfer coefficient is h. Three cases are shown in Figure 7.3, for which the equations to be satisfied by relaxation are given below.

Plane surface (a) $\quad \dfrac{T_1}{2} + \dfrac{T_3}{2} + T_2 + \left(\dfrac{ha}{k}\right) T_f - \left(2 + \dfrac{ha}{k}\right) T_0 = 0$

External corner (b) $\quad \dfrac{T_1}{2} + \dfrac{T_2}{2} + \left(\dfrac{ha}{k}\right) T_f - \left(1 + \dfrac{ha}{k}\right) T_0 = 0$

Internal corner (c) $\quad \dfrac{T_1}{2} + \dfrac{T_2}{2} + T_3 + T_4 + \left(\dfrac{ha}{k}\right) T_f - \left(3 + \dfrac{ha}{k}\right) T_0 = 0$

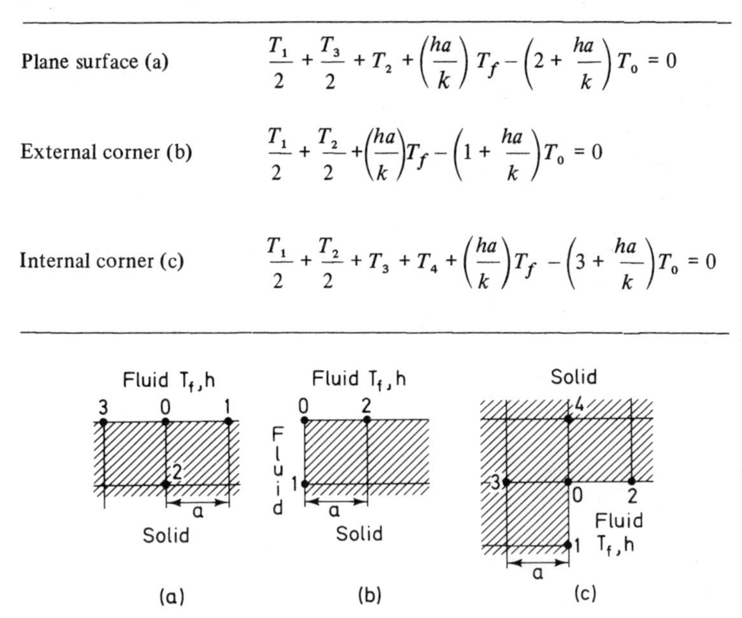

Figure 7.3 Convection effects at boundaries

The heat transfer rate between the wall and the fluid is given by

$$\dot{Q} = \Sigma h\,a\,(T_m - T_f) \text{ per unit height} \tag{7.11}$$

where T_f is the fluid temperature and T_m is the temperature at each *surface* mesh point.

If $\dot{Q}'''$ is not zero (electric current-carrying conductor), the residual is adjusted to $(-\dot{Q}'''a^2/k)$ rather than zero.

These techniques can be extended to deal with angled and curved boundaries, radiation at the surface, cylindrical boundaries (using cylindrical coordinates (r, z, ϕ) and for *transient* heat transfer when $(\partial T/\partial \tau)$ is not zero.[1,2]

7.8 Surface heat transfer coefficient

Analysis of convection in a boundary layer is complex and dimensional analysis or analogy techniques are often used. Convection may be *natural* or *free*, when the fluid motion is due to buoyancy forces generated by temperature differences, or *forced* when the flow is pumped. In either case the flow pattern may be *laminar* or *turbulent*. The range of values of h depends on the type of convection:

Forced convection:	gases	10–700 W/m²K
	liquids	100–10 000 W/m²K
Free convection:	gases	0.5–500 W/m²K
	liquids	50–2000 W/m²K

For *forced convection* it is found by dimensional analysis that

$$\text{Nu} = \phi\,(\text{Re}, \text{Pr}) = \text{constant Re}^a\,\text{Pr}^b \tag{7.12}$$

where Nu is the *Nusselt* number $\left(\dfrac{hl}{k}\right)$

Pr is the *Prandtl* number $\left(\dfrac{\mu C_p}{k}\right)$

and Re is the *Reynolds* number $\left(\dfrac{\rho Vl}{\mu}\right)$

In these dimensionless groups k is the thermal conductivity of the fluid, C_p is the specific heat capacity of the fluid, ρ is the fluid density, μ is the fluid viscosity, V is the fluid free stream or bulk velocity and l is a representative length dimension. For a plate, the accepted representative length is the distance from the leading edge x, and for a tube the diameter d is used.

The surface heat transfer coefficient is contained in the Nusselt number, and to enable h to be determined experiments are necessary to find the values of the equation constant and the indices a and b for any particular heat exchange surface.

The criterion which decides whether flow is laminar or turbulent in forced convection is the value of the Reynolds number:

for a plane surface	$Re < 500\,000$ flow is laminar
	$Re > 500\,000$ flow is turbulent
and for a tube	$Re < 2300$ flow is laminar
	$Re > 2300$ flow is turbulent

The fluid properties μ, k, ρ, and C_p fall into two groups. Bulk properties ρ and C_p, which are concerned with the enthalpy of the whole flow of fluid are usually evaluated at the bulk or free stream temperature whereas transport properties, μ and k, which are concerned with momentum and energy transfer in regions of velocity and temperature gradient are sometimes evaluated at the film temperature T_f. Film temperature is the arithmetic mean between the bulk or free stream temperature, T_b and the wall temperature T_w

$$T_f = \frac{T_b + T_w}{2}$$

When using any empirical equation care must be taken to use the prescribed temperatures to evaluate the fluid properties. Tables of fluid properties needed for convective heat transfer for air and water are included with the programs in this chapter.

Typical values for the constant and indices a and b are tabulated below for plane surfaces and tubes with laminar and turbulent flow. These relations will give *average* Nusselt numbers for a finite length of surface. Fluid properties should be evaluated at temperatures which are an average for the length considered. If results for other configurations are required reference should be made to specialised texts.[1,2]

Surface	*Laminar flow*	*Turbulent flow*	*Properties at*
Plane	$Nu = 0.664 Re^{0.5} Pr^{0.33}$	$Nu = 0.036 Re^{0.8} Pr^{0.33}$	Film temperature
Tube	$Nu = 3.66$ (long tube)	$Nu = 0.023 Re^{0.8} Pr^{0.4}$	Bulk temperature

As an alternative to dimensional analysis an analogy may be used. The simplest relation is *Reynolds'* analogy modified by *Colburn*. This states, for a plane surface with laminar or turbulent flow and for turbulent flow in a tube, that

$$St Pr^{2/3} = \frac{f}{2} \quad \text{with } 0.5 < Pr < 100 \tag{7.13}$$

where St is the Stanton number

$$\frac{Nu}{Re Pr} = \frac{h}{\rho V C_p}$$

and f is the friction factor obtained by experiment or, for *smooth* pipes, by Blasius' equation ($f = 0.079\text{Re}^{-1/4}$). It can be seen that h is contained in the Stanton number.

In *free convection* it is found by dimensional analysis

$$\text{Nu} = \phi(\text{Pr}, \text{Gr}) \tag{7.14}$$

where Gr is the *Grashof* number

$$\left(\frac{\rho^2 \beta g \theta l^3}{\mu^2}\right)$$

β is the coefficient of cubical expansion ($\beta = 1/T$ for a perfect gas), g is the gravitational acceleration and θ a temperature difference.

The product of Pr and Gr is called the *Rayleigh* number, Ra.

$$\text{Ra} = \text{Pr.Gr} = \frac{\rho^2 l^3 C_p \beta g \theta}{k\mu}$$

It is found that the onset of turbulence in free convection is determined by the numerical value of the Rayleigh number in a similar way that the Reynolds number determined the nature of forced convection flow.

It is found by experiment for vertical plane or cylindrical surfaces that when $\text{Ra} < 10^9$, flow is laminar and $\text{Nu} = 0.59\,(\text{PrGr})^{1/4}$ whereas for $\text{Ra} > 10^9$, flow is turbulent and $\text{Nu} = 0.13(\text{PrGr})^{1/3}$). For these cases, the representative length chosen for Gr is height and the resulting Nu are mean values over the whole surface. Film temperature is used for properties. If results for other configurations are required reference should be made to specialist texts. [1,2]

7.9 Heat exchangers

A heat exchanger consists of a solid, good conducting boundary separating two fluids which are exchanging energy by heat transfer. The wall is thin so that its thermal resistance is negligible (unless scale covered) and the overall heat transfer coefficient is given by

$$\frac{1}{U} = \sum \frac{1}{h} \text{ or } \frac{1}{U'} = \sum \frac{1}{2\pi r h}$$

Figure 7.4 shows that the temperature difference θ between the two fluid streams varies along the length of the heat exchanger and some mean θ is required. The variation is not linear and it is found that the *log mean temperature difference* (LMTD) is needed.

$$\theta_{\text{LMTD}} = \frac{\theta_1 - \theta_2}{\ln(\theta_1/\theta_2)} \quad (7.15)$$

where θ_1 and θ_2 are the temperature differences at either end.

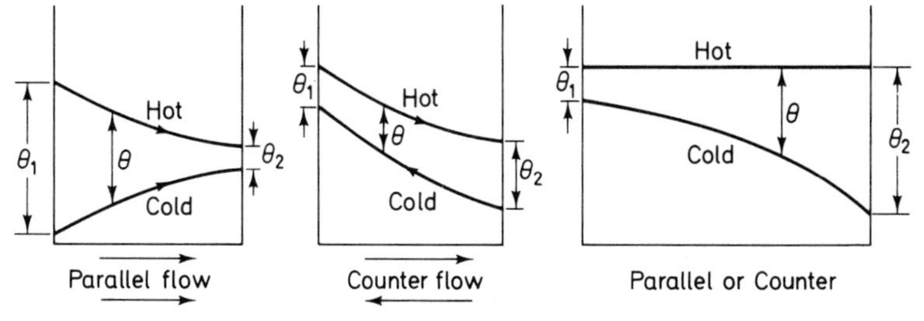

Figure 7.4 Heat exchanger temperature profiles

The heat transfer problem is solved by using

$$\dot{Q} = UA\,\theta_{\text{LMTD}} \quad \text{or} \quad \dot{Q} = U'L\,\theta_{\text{LMTD}} \quad (7.16)$$

together with the steady flow energy equation applied along each stream

$$\dot{Q} = \dot{m}_H C_{p_H} \Delta T_H = \dot{m}_c C_{p_c} \Delta T_c \quad (7.17)$$

The value of θ_{LMTD} is greater in *counter* flow exchangers which should always be used in preference to *parallel* flow designs since the heat transfer area, i.e. the exchanger size, will be least. (When a phase change is involved, there is no difference.)

7.10 NTU method for heat exchangers

Design of a heat exchanger is not easy by the method of section 7.9 since temperatures and U values cannot be determined unless assumptions about size and velocity are made, followed by repetitive calculations and amendments. Some problems are avoided by the NTU method. (NTU is the symbol for the 'number of transfer units' associated with the heat exchanger.)

Let *capacity ratio*, $C = \dfrac{\dot{m}_c C_{p_c}}{\dot{m}_H C_{p_H}}$

or $\dfrac{\dot{m}_H C_{p_H}}{\dot{m}_c C_{p_c}}$ (7.18)

such that $C < 1$

Let *effectiveness*, $E = \dfrac{\text{actual heat transfer}}{\text{maximum possible heat transfer}}$

(The maximum possible heat transfer will occur when $t_{H_2} = t_{c_1}$ or $t_{c_2} = t_{H_1}$, which being determined by whether $\dot{m}_H C_{pH} > \dot{m}_c C_{pc}$ or $\dot{m}_c C_{pc} > \dot{m}_H C_{pH}$)

If, for example, (Figure 7.4) $\dot{m}_H C_{pH} > \dot{m}_c C_{pc}$,

$$E = \frac{t_{c_2} - t_{c_1}}{t_{H_1} - t_{c_1}}, \quad \text{i.e. } t_{c_2} \rightarrow t_{H_1}$$

Let the *number of transfer units,* $\text{NTU} = \dfrac{UA}{\dot{m}_c C_{p_c}}$

or $$\frac{U'L}{\dot{m}_c C_{p_c}} \quad \text{if } \dot{m}_H C_{pH} > \dot{m}_c C_{p_c}$$

and $$\text{NTU} = \frac{UA}{\dot{m}_H C_{pH}}$$

or $$\frac{U'L}{\dot{m}_H C_{pH}} \quad \text{if } \dot{m}_c C_{p_c} > \dot{m}_H C_{pH}$$

(The denominator is the smaller capacity)

Then in *counter* flow it is found that

$$E = \frac{1 - e^{-\text{NTU}(1-C)}}{1 - Ce^{-\text{NTU}(1-C)}} \qquad (7.19)$$

This relation can be presented on a graph and graphs are also available for other common designs.[3]

7.11 Fins

With given values of surface heat transfer coefficients and a given value of the mean temperature difference between two streams, the only way of increasing the heat transfer rate is to increase the surface area. This can be achieved by the use of correctly placed extended surfaces such as fins, rather than by an overall increase in size.

In any heat transfer problem, there may be one dominant resistance. In the case of a liquid flowing along a thin tube and being cooled by heat transfer to air outside the tube, the surface heat transfer coefficient on the liquid side of the tube is very much greater than that on the air side of the tube. The main resistance to heat transfer is therefore on the air surface, and the heat transfer rate can be enhanced by putting fins on this surface. There would be no point in putting fins on the liquid side.

The addition of a fin may affect the surface heat transfer coefficient as the flow pattern will be altered and for complex finned heat exchangers diagrams are available[3] to determine the value of h. The diagrams also include data for f (the friction factor) to determine the pressure loss in the heat exchanger matrix.

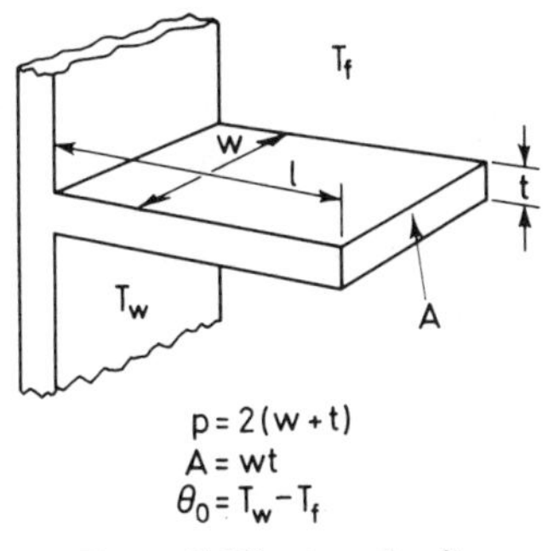

Figure 7.5 Rectangular fin

In any fin the temperature changes from the wall temperature at its root to a temperature approaching that of the surrounding fluid at the tip. A simple analysis for slender circular or rectangular fins can be made giving the temperature difference θ at distance x from the root as (at the root, $\theta = \theta_{wall}$)

$$\frac{\theta}{\theta_{wall}} = \frac{\cosh m(l - x)}{\cosh ml} \tag{7.20}$$

together with the heat transfer rate associated with the fin (Figure 7.5)

$$\dot{Q} = m\, k\, A\, \theta_{wall} \tanh ml \tag{7.21}$$

where $m = \left(\frac{ph}{kA}\right)^{1/2}$

p is the fin perimeter
l is the fin length
k is the thermal conductivity of the fin
A is the fin cross-sectional area
h is the average surface heat transfer coefficient

and θ_{wall} is the temperature difference between wall and fluid.

A fin efficiency is then defined by

$$\eta_{fin} = \frac{\text{actual heat transfer rate through the fin}}{\text{heat transfer rate if the whole fin were at the wall temperature}}$$

$$\eta_{fin} = \frac{\tanh ml}{ml}$$

If a more detailed analysis is made the result is often no more accurate since the value of h will be an estimated average value for the whole fin area. For cylindrical and wedge-shaped fins the equations are not easily solved and graphs are available[1] to determine fin performance, in terms of fin efficiency.

7.12 Radiation

A surface at some temperature T that is within sight of some other surface will both emit and receive radiation. The heat transfer rate will be the algebraic sum of these two quantities. The radiation flux $\dot{E}$ is defined as the time rate of radiation energy crossing a system boundary. The ***emissive power*** of a surface is the radiation flux per unit area $\dot{E}''$.

Incident radiation may be reflected, absorbed or transmitted. If ρ = reflectivity, τ = transmissivity and a = absorptivity then

$$\rho + \tau + a = 1$$

For solids and liquids, it may usually be assumed that $\tau = 0$ and, for gases, $\rho = 0$. In the case of a gas between two solids exchanging radiation it is often assumed that $\tau = 1$.

For simple analysis, a perfect emitter and receiver is required. A *black body* is defined as a body with a surface whose absorptivity is unity under all conditions. Such a body will absorb all incident radiation. The *Stefan–Boltzmann* law states that the emissive power of a black body is proportional to the fourth power of thermodynamic temperature

$$\dot{E}''_b = \sigma T^4 \tag{7.22}$$

where σ is the Stefan–Boltzmann constant, equal to 5.67×10^{-8} W/m^2 K^4. The radiation is spread over a wide spectrum of wavelength or frequency, governed by Planck's law (Figure 7.6).

Real surfaces are not black and they emit less radiation than a black surface. The monochromatic ***emissivity***, ϵ_λ of a surface, is the ratio of

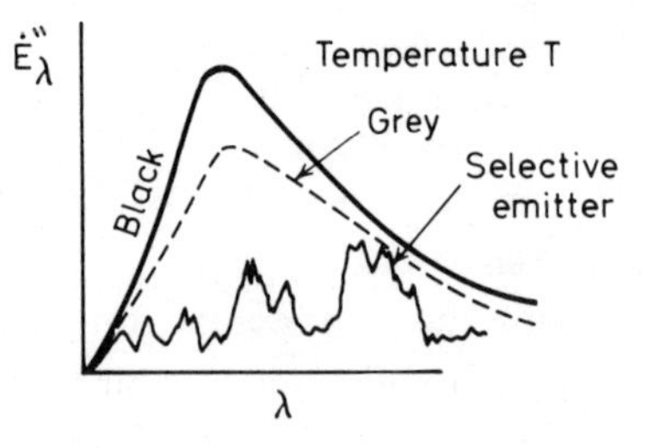

Figure 7.6 Radiation spectrum

the emissive power of a surface at wavelength λ to the emissive power of a black body at the same wavelength and at the same temperature T.

$$\epsilon_\lambda = \left(\frac{\dot{E}''_\lambda}{\dot{E}''_{b\lambda}}\right)_T$$

A *grey body* has an emissivity which is independent of temperature and wavelength. For this body ϵ = constant (Figure 7.6).

$$\epsilon = \left(\frac{\dot{E}''}{\dot{E}''_b}\right) \tag{7.23}$$

The monochromatic emissivity of a real surface varies with wavelength and temperature in some arbitrary way and Figure 7.6 illustrates how the emissivity of a real body called a *selective emitter* might vary.

Kirchoff's law states that the emissivity of a grey surface at temperature T is equal to its absorptivity (for radiation received from another surface) at the same temperature T.

7.13 Grey bodies enclosed in black surroundings

The heat transfer rate to a body will be the difference between the energy absorbed and the energy emitted. The heat transfer rate to the

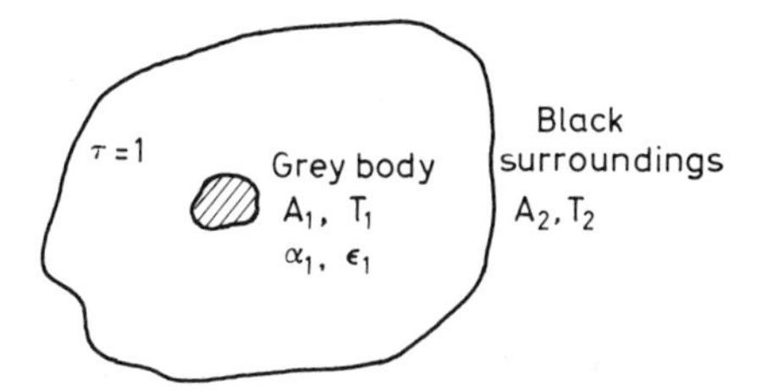

Figure 7.7 Grey body in black surroundings

grey body at temperature T_1, completely surrounded (Figure 7.7) by the black surface at temperature T_2, is given by

$$\dot{Q} = \epsilon_1 \sigma A_1 (T_2^4 - T_1^4) \tag{7.24}$$

This result enables a useful approximation to be made for real bodies. If a representative value of ϵ is used, then the real body may be treated as a grey body, and the equation may be applied to a grey body in large surroundings. If the surroundings are not black, some of the incident energy will be reflected, but as the surroundings are large compared with the body, it is unlikely that a significant amount of the reflected radiation will be re-incident on the body to invalidate the result.

It is sometimes convenient, in such simple cases when there is more than one mode of heat transfer, to use a radiation heat transfer coefficient, h_r such that

$$\dot{Q}''_{rad} = h_r\theta$$

where θ is the temperature difference between radiator and receiver. This may be added to the surface heat transfer coefficient for convection. In this case

$$h_r = \epsilon_1 \sigma(T_1 + T_2)(T_1^2 + T_2^2)$$

7.14 Radiation heat transfer for unenclosed surfaces

In order to pursue more general heat transfer problems than the grey body entirely surrounded by a receiving surface it is necessary to study the spatial distribution of the energy radiation by a small surface (Figure 7.8).

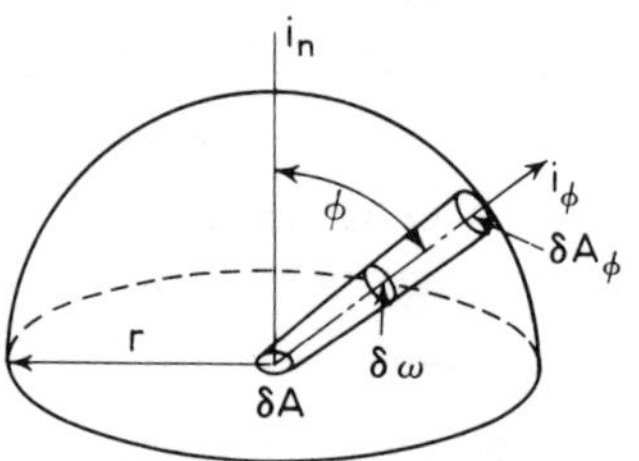

Figure 7.8 Intensity of radiation

Lambert's law of diffuse radiation states that in any direction ϕ the intensity of radiation $i_\phi = i_n \cos\phi$ where i_n is the intensity of radiation in a direction normal to the area.

For a black surface

$$i_n = \frac{\sigma T^4}{\pi}$$

and for a grey surface

$$i_n = \frac{\epsilon\sigma T^4}{\pi}$$

The heat transfer rate from one radiating surface to another will depend on three factors:

(i) the amount that each surface can see of the other surface ($i_{\phi 1}$ and $i_{\phi 2}$);

(ii) the emissivity of the surfaces (ϵ_1 and ϵ_2);

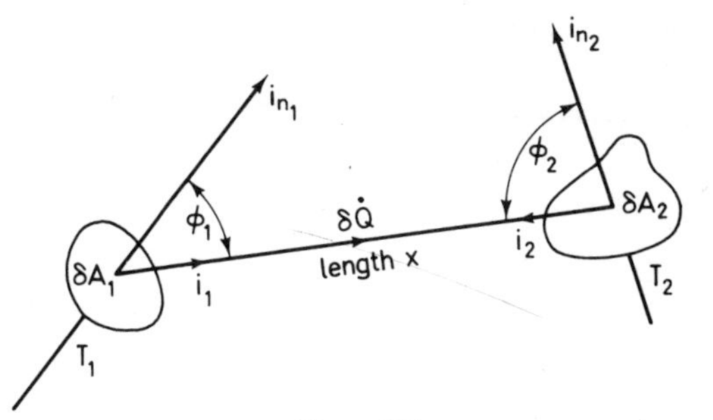

Figure 7.9

(iii) the temperature of the surfaces, (T_1 and T_2).

Consider two arbitrarily disposed grey surfaces shown in Figure 7.9. The heat transfer rate from surface 1 to surface 2 is given by:

(the rate of energy emitted and reflected by 1 and intercepted and absorbed by 2 less the rate of energy emitted and reflected by 2 and intercepted and absorbed by 1).

This may be written

$$\dot{Q}_{12} = F_E F_G \sigma A (T_1^4 - T_2^4) \tag{7.25}$$

where F_G is a geometric factor and F_E is an emissivity factor.

The geometric factor F_G is defined by: F_G = the fraction of energy emitted per unit time by one surface that is intercepted by the other surface.

It can be shown that for surfaces 1 and 2

$$F_{12} = \frac{1}{A_1} \int_{A_1} \int_{A_2} \frac{\cos \phi_1 \cos \phi_2}{\pi x^2} \, dA_1 \, dA_2$$

and

$$F_{21} = \frac{1}{A_2} \int_{A_2} \int_{A_1} \frac{\cos \phi_1 \cos \phi_2}{\pi x^2} \, dA_2 \, dA_1$$

and it can be seen that

$$A_1 F_{12} = A_2 F_{21} = \int_{A_1} \int_{A_2} \frac{\cos \phi_1 \cos \phi_2}{\pi x^2} \, dA_1 \, dA_2 \tag{7.26}$$

(F_{12} means the fraction of energy leaving surface 1 that is intercepted by surface 2). Geometric factors are found by direct integration for simple cases or from *Hottel* charts[4] (Figure 7.10) for situations that occur frequently.

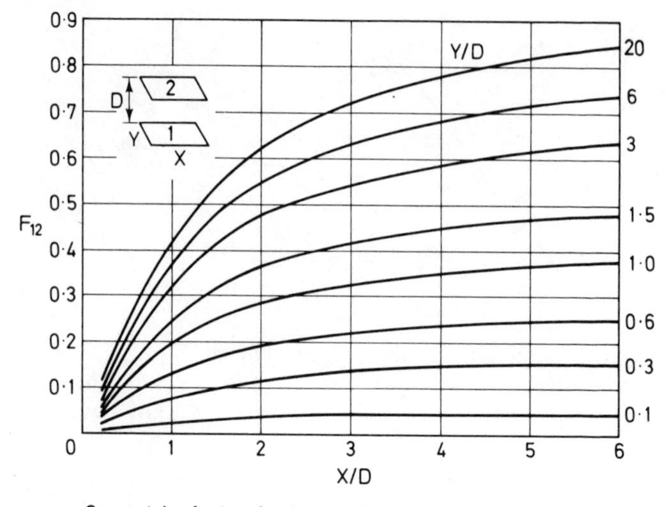

Geometric factor for two rectangular parallel plates

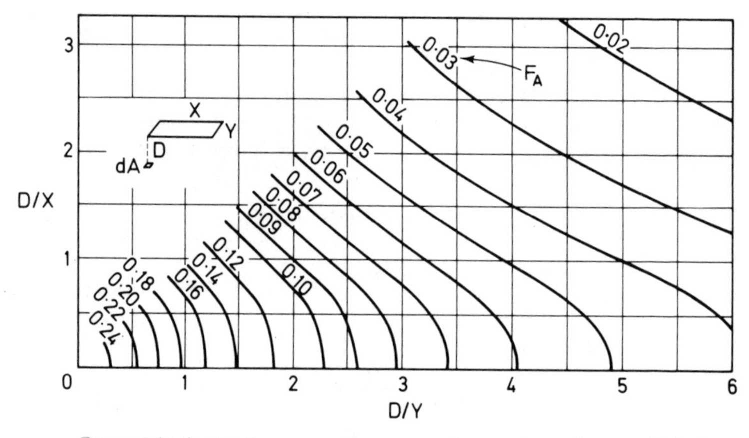

Geometric factor for a small area and a rectangular parallel plate

Figure 7.10 Geometric factor (Hottel) charts

It is possible to deduce the emissivity factor analytically for cases where the geometric factor is easily determined. The factor for two infinite parallel plane surfaces (for which $F_G = 1$) is obtained by summing an infinite series obtained by considering successive reflections from each surface in turn. In this case the emissivity factor F_E is found to be

$$F_E = \left(\frac{1}{\frac{1}{\epsilon_1} + \frac{1}{\epsilon_2} - 1} \right)$$

where ϵ_1 and ϵ_2 are the emissivities of the grey surfaces 1 and 2.

Similarly, for two long coaxial cylinders the emissivity factor F_E is given by

$$\frac{1}{F_E} = \left\{ \frac{1}{\epsilon_1} + \frac{d_1}{d_2} \left(\frac{1}{\epsilon_2} - 1 \right) \right\}$$

where ϵ is the emissivity, d the diameter, and subscripts 1 and 2 refer to the inner and outer cylinders respectively.

7.15 Electrical analogy for grey body radiation

For a more general grey body case, when F_G is not equal to unity, an electrical analogy may be used. Each surface is represented by a black body potential $\dot{E}_b''$. This potential is reduced by a resistance $(\rho/A\epsilon)$ to become the grey body potential $\dot{J}''$. The geometric factor resistance $(1/AF_G)$ between the surfaces is then placed between the grey body potentials to form the network.

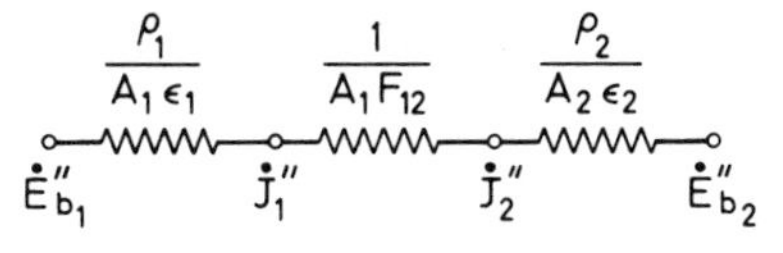

Figure 7.11 Radiation resistance network

Figure 7.11 shows a resistance network for two grey bodies. F_{12} is obtained from the Hottel chart (or by calculation). Then assuming that the medium between the two surfaces has $\tau = 1$, so that $\rho_1 = 1 - \epsilon_1$ and $\rho_2 = 1 - \epsilon_2$, the electrical analogy has a potential difference

$$V = \dot{E}''_{b_1} - \dot{E}''_{b_2} = \sigma (T_1^4 - T_2^4)$$

and a total resistance

$$R = \frac{\rho_1}{A_1 \epsilon_1} + \frac{1}{A_1 F_{12}} + \frac{\rho_2}{A_2 \epsilon_2}$$

so that $\dot{Q} = I = V/R$

$$= \frac{\sigma(T_1^4 - T_2^4)}{\frac{\rho_1}{A_1 \epsilon_1} + \frac{1}{A_1 F_{12}} + \frac{\rho_2}{A_2 \epsilon_2}} \qquad (7.27)$$

The 'circuit diagram' becomes more involved for multiple surface problems when each one can see some or all of the others.

This method will also be satisfactory if one or more of the surfaces can be considered black. In such a case, $\epsilon = 1$ for a black surface and there is no resistance so that the black body potential and the grey body potential are equal for the surface.

7.16 References

1. Holman, J.P., *Heat Transfer,* McGraw-Hill, New York (1972).
2. Adams, J.A. and Rogers, D.F., *Computer Aided Heat Transfer Analysis,* McGraw-Hill, New York (1973).
3. Kays, W.M. and London, A.L., *Compact Heat Exchangers,* McGraw-Hill, London (1964).
4. Hamilton, D.C. and Morgan, W.R., *Radiant Interchange Configuration Factors,* N.A.C.A. TN-2386 (1952).

WORKED EXAMPLES

Heat transfer data

```
LINES 2310 TO 2400 FOR AIR,TEMPERATURE RANGE -100 C TO 800 C
LINES 2410 TO 2600 FOR WATER,TEMPERATURE RANGE 0.01 C TO 325 C

EACH LINE CONTAINS DATA FOR TEMPERATURE (T) IN C,SPECIFIC VOLUME(VS) IN M↑3/KG,
SPECIFIC HEAT AT CONSTANT PRESSURE(CP) IN KJ/KG K,THERMAL CONDUCTIVITY(K)
IN W/M K,DYNAMIC VISCOSITY(MU) IN G/S M AND PRANDTL NUMBER(PR).
LIST
 2310  DATA -100,0.488,1.01,0.016,0.012,0.75
 2320  DATA 0,0.773,1.01,0.024,0.017,0.72
 2330  DATA 100,1.057,1.02,0.032,0.022,0.7
 2340  DATA 200,1.341,1.03,0.039,0.026,0.69
 2350  DATA 300,1.624,1.05,0.045,0.03,0.69
 2360  DATA 400,1.908,1.07,0.051,0.033,0.7
 2370  DATA 500,2.191,1.1,0.056,0.036,0.7
 2380  DATA 600,2.473,1.12,0.061,0.039,0.71
 2390  DATA 700,2.756,1.14,0.066,0.042,0.72
 2400  DATA 800,3.039,1.16,0.071,0.044,0.73

2410LIST
 2410  DATA 0.01,0.001,4.217,0.569,1.755,13.02
 2420  DATA 10,0.001,4.193,0.587,1.301,9.29
 2430  DATA 20,0.001,4.182,0.603,1.002,6.95
 2440  DATA 30,0.001,4.179,0.618,0.797,5.39
 2450  DATA 40,0.00101,4.179,0.632,0.651,4.31
 2460  DATA 50,0.00101,4.181,0.643,0.544,3.53
 2470  DATA 60,0.00102,4.185,0.653,0.462,2.96
 2480  DATA 70,0.00102,4.19,0.662,0.4,2.53
 2490  DATA 80,0.00103,4.197,0.67,0.35,2.19
 2500  DATA 90,0.00104,4.205,0.676,0.311,1.93
 2510  DATA 100,0.00104,4.216,0.681,0.278,1.723
 2520  DATA 125,0.00107,4.254,0.687,0.219,1.358
 2530  DATA 150,0.00109,4.31,0.687,0.18,1.133
 2540  DATA 175,0.00112,4.389,0.679,0.153,0.99
 2550  DATA 200,0.00116,4.497,0.665,0.133,0.902
 2560  DATA 225,0.0012,4.648,0.644,0.1182,0.853
 2570  DATA 250,0.00125,4.867,0.616,0.1065,0.841
 2580  DATA 275,0.00132,5.202,0.582,0.0972,0.869
 2590  DATA 300,0.0014,5.762,0.541,0.0897,0.955
 2600  DATA 325,0.00153,6.861,0.493,0.079,1.1
```

Note: In the examples using data from these blocks only one (7.6) has used an interpolation routine as a demonstration. In the remainder the data has been obtained by inspection and inserted into the programs directly. This method has been used since the data is constant for each problem and it is therefore more convenient to use simple direct insertion rather than a lengthy interpolation routine.

Example 7.1 U values

The walls of a house are often made up of two layers of solid material separated by an air gap. There is convection at each exposed outer surface and the air gap is given a thermal resistance to allow for all heat transfers within the gap.

Write a program to determine the heat transfer rate through the walls of a house from the inside air to the outside air. Evaluate the program for the following cases for which conductivity and resistance data is given:

(a) a cavity wall with a 105 mm brick outer layer for which $U = 3.3$ W/m^2 K, a ventilated air gap 36 mm wide for which $R = 0.18$ m^2 K/W and an inner layer of brick 105 mm thick with 16 mm plaster for which $U = 2.7$ W/m^2 K;

(b) the cavity wall of (a) with the air gap filled with an injected insulating material with thermal conductivity $k = 0.036$ W/m K.

For both cases the surface resistance on the inside wall is 0.2 m^2 K/W and on the outside wall (in normal exposure and winds) is 0.06 m^2 K/W. Wall area, temperature difference, heating weeks and hours, fuel cost and capital cost may be chosen to suit any particular problem.

The program should determine the effectiveness of the cavity insulation in terms of payback time for the filling, ignoring future price rises in fuel. It should also be capable of altering the wall specification and resistance. Values of resistance for various wall constructions may be found in *IHVE Guide,* Volume A, Institution of Heating and Ventilating Engineers, London (1971).

The necessary reading for this example is in section 7.5.

```
LIST
 10  PRINT "EXAMPLE 7.1"
 20  PRINT
 30  PRINT "WALL DATA"
 40  PRINT "WHAT IS THE OUTSIDE SURFACE RESISTANCE?"
 50  INPUT SO
 60  PRINT "WHAT IS THE INSIDE SURFACE RESISTANCE?"
 70  INPUT SI
 80  PRINT "WHAT IS THE INNER SKIN U VALUE?"
 90  INPUT D
 100  PRINT "WHAT IS THE OUTER SKIN U VALUE?"
```

```
110  INPUT E
120  PRINT "WHAT IS THE CAVITY RESISTANCE?"
130  REM IF FILLED R=THICKNESS/THERMAL CONDUCTIVITY,IF SOLID R=0
140  INPUT RC
150  RI=1/D
160  RO=1/E
170  R=SO+SI+RI+RO+RC
180  U=1/R
190  PRINT "WHAT IS THE WALL AREA?"
200  INPUT A
210  PRINT "WHAT IS THE TEMPERATURE DIFFERENCE?"
220  INPUT T
230  PRINT "HOW MANY HEATING WEEKS PER ANNUM?"
240  INPUT W
250  PRINT "HOW MANY HEATING HOURS PER DAY?"
260  INPUT H
270  PRINT "WHAT IS THE ENERGY COST IN POUNDS PER KWH?"
280  INPUT C
290  HL=U*A*T
300  PRINT "HEAT LOSS RATE";HL;"W"
310  TL=(HL*7*W*H*C)/1000
320  PRINT "HEATING COST IN POUNDS?";TL
330  PRINT "DO YOU WANT ANOTHER WALL? IF YES,TYPE 1; IF NO, TYPE 2"
340  INPUT X
350  IF X=1 THEN GOTO 30
360  PRINT "IS THERE A SAVINGS CALCULATION? IF YES, TYPE 1; IF NO,TYPE 2"
370  INPUT Y
380  IF Y=2 THEN GOTO 470
390  PRINT "TO FIND SAVINGS SUBTRACT THE COSTS;DATA TL VALUES ABOVE"
400  INPUT T1,T2
410  S=T1-T2
420  PRINT "WHAT IS CAPITAL COST IN POUNDS?"
430  INPUT CAP
440  REM NO INFLATION ALLOWANCE IN FUEL COST
450  PBT=CAP/S
460  PRINT "PAYBACK TIME WITHOUT INFLATION ALLOWANCE";PBT;"YEARS"
470  END
```

```
RUN
EXAMPLE 7.1

WALL DATA
WHAT IS THE OUTSIDE SURFACE RESISTANCE?
? .06
WHAT IS THE INSIDE SURFACE RESISTANCE?
? .2
WHAT IS THE INNER SKIN U VALUE?
? 2.7
WHAT IS THE OUTER SKIN U VALUE?
? 3.3
WHAT IS THE CAVITY RESISTANCE?
? .18
WHAT IS THE WALL AREA?
? 180
WHAT IS THE TEMPERATURE DIFFERENCE?
? 15
HOW MANY HEATING WEEKS PER ANNUM?
? 30
HOW MANY HEATING HOURS PER DAY?
? 12
WHAT IS THE ENERGY COST IN POUNDS PER KWH?
? .04
HEAT LOSS RATE 2425.003W
HEATING COST IN POUNDS? 244.4403
DO YOU WANT ANOTHER WALL? IF YES,TYPE 1; IF NO. TYPE 2
? 1
WALL DATA
WHAT IS THE OUTSIDE SURFACE RESISTANCE?
? .06
WHAT IS THE INSIDE SURFACE RESISTANCE?
? .2
```

```
WHAT IS THE INNER SKIN U VALUE?
? 2.7
WHAT IS THE OUTER SKIN U VALUE?
? 3.3
WHAT IS THE CAVITY RESISTANCE?
? 1
WHAT IS THE WALL AREA?
? 180
WHAT IS THE TEMPERATURE DIFFERENCE?
? 15
HOW MANY HEATING WEEKS PER ANNUM?
? 30
HOW MANY HEATING HOURS PER DAY?
? 12
WHAT IS THE ENERGY COST IN POUNDS PER KWH?
? .04
HEAT LOSS RATE 1396.503W
HEATING COST IN POUNDS? 140.767511
DO YOU WANT ANOTHER WALL? IF YES,TYPE 1; IF NO, TYPE 2
? 2
IS THERE A SAVINGS CALCULATION? IF YES, TYPE 1; IF NO,TYPE 2
? 1
TO FIND SAVINGS SUBTRACT THE COSTS;DATA TL VALUES ABOVE
? 244.4403? 140.7675
WHAT IS CAPITAL COST IN POUNDS?
? 500
PAYBACK TIME WITHOUT INFLATION ALLOWANCE 4.82286579YEARS

STOP AT 470
```

Program nomenclature

RI	inner skin resistance
RO	outer skin resistance
R	overall resistance
HL	heat loss rate
TL	heating costs
T1	heating costs
T2	heating costs
S	cost savings
PBT	payback time

Program notes

(1) In lines 210 to 260 the program inputs assume that the whole house envelope is kept 15°C higher than the mean ambient temperature of 5°C and that heating is used 12 hours a day for 30 weeks per annum. If less heating is required the payback period would increase but fuel cost inflation would (if included) reduce the period.
(2) Lines 30 to 140 give input data which can be changed for different building materials.
(3) Lines 150 to 180 calculate the *U* value for the wall.
(4) By repeating the calculation (line 350) the savings achieved by insulation can be determined and divided into the capital cost to obtain

a simple payback calculation. For large scale costs the effects of interest on capital, inflation and depreciation would need to be considered by proper accounting techniques but for household calculations the simple method should suffice.
(5) Problem 7.1 extends the work to include flooring, loft insulation and double glazing, and by suitable choice of areas and temperature differences could be applied to terraced, semi-detached or detached houses, flats or bungalows.

Example 7.2 Two-dimensional conduction

Figure 7.12 shows a chimney of rectangular cross-section for which the wall temperatures and thermal conductivity are known. Write a program to determine the temperature distribution in the chimney

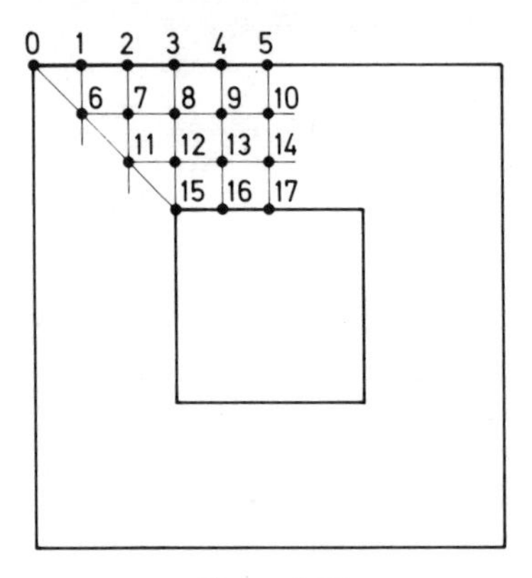

Figure 7.12

cross-section and the heat transfer rate per unit height. The relaxation grid size should be chosen to give two points between inside and outside surfaces. The problem solution will need only 1/8 of the whole cross-section since it is symmetrical.

Test data

Chimney dimensions: outside 2 m × 2 m, inside 0.8 m × 0.8 m
Inner wall temperature 225°C
Outer wall temperature 25°C
Thermal conductivity of chimney bricks 0.9 W/m K
The mesh points are labelled in a convenient manner. At any position in the cross-section the equation to be satisfied is

$$\frac{\partial^2 T}{\partial x^2} + \frac{\partial^2 T}{\partial y^2} = 0$$

which requires that, for example,

$$T(8) = [T(3) + T(9) + T(12) + T(7)]/4$$

Not all points will have the same equation; for example,

$$T(6) = [2\ T(1) + 2\ T(7)]/4$$

To solve this problem each mesh point is given an initial temperature some of these are the given boundary conditions but the rest are arbitrary (100°C). The nine mesh points with arbitrary values are then solved to give more accurate temperatures which may be printed if required. Successive solutions are made until the change in the temperatures achieved is considered negligible.

Experience assists in knowing how many times it is worth repeating the exercise. In this program any number of iterations can be made and printed if required. The program could be modified to run until the difference in successive values of temperature was within some chosen limit before printing the results.

As an alternative to the iterative method it is possible to solve the nine equations simultaneously by numerical techniques.

The necessary reading for this example is in section 7.7.

```
LIST
 10  PRINT "EXAMPLE 7.2"
 20  PRINT
 30  DIM T[18]
 40  FOR N=0 TO 17
 50   READ T[N]
 60  NEXT N
 70  PRINT "THE 9 POINTS TO RELAX ARE 6 TO 14"
 80  PRINT "POINTS 7,8 AND 9 ARE THE SAME,12 AND 13 ARE SAME,"
 85  PRINT "6,10,11 AND 14 ARE SINGULAR"
 90  PRINT
 100  A=0
 110  N=6
 120  T[N]=(2*T[1]+2*T[7])/4
 130  FOR N=7 TO 9
 140   T[N]=(T[N-5]+T[N+4]+T[N+1]+T[N-1])/4
 150  NEXT N
 160  N=10
 170  T[N]=(T[5]+T[14]+2*T[9])/4
 180  N=11
 190  T[N]=(2*T[7]+2*T[12])/4
 200  FOR N=12 TO 13
 210   T[N]=(T[N-4]+T[N+3]+T[N+1]+T[N-1])/4
 220  NEXT N
 230  N=14
 240  T[N]=(T[10]+T[17]+2*T[13])/4
 250  A=A+1
 260  PRINT "NUMBER OF RELAXATIONS";A
 270  PRINT
 280  PRINT "DO YOU REQUIRE A PRINT OUT?IF YES,TYPE 1;NO, TYPE 2"
 290  INPUT Z
 300  IF Z=2 THEN GOTO 110
 310  PRINT "POINT NUMBER  TEMPERATURE"
 320  PRINT
 330  FOR N=0 TO 17
 340   PRINT N,T[N]
 350  NEXT N
 360  PRINT "DO YOU WISH ANOTHER RELAXATION?IF YES,TYPE 1;NO,TYPE 2"
```

```
370  INPUT X
380  IF X=1 THEN GOTO 110
390  REM FIND HEAT TRANSFER RATE AT INSIDE AND OUTSIDE
400  QIN=(4*(225-T[14])+8*(225-T[13])+8*(225-T[12]))*0.9
410  QOT=(4*(T[10]-25)+8*((T[9]-25)+(T[8]-25)+(T[7]-25)+(T[6]-25)))*0.9
420  PRINT
430  PRINT "HEAT TRANSFER RATE TO CHIMNEY=";QIN;"W/M"
440  PRINT
450  PRINT "HEAT TRANSFER RATE FROM CHIMNEY=";QOT;"W/M"
460  END
470  DATA 25,25,25,25,25,25,100,100,100,100,100,100.100,100,100,225,225,225
```

```
RUN
EXAMPLE 7.2

THE 9 POINTS TO RELAX ARE 6 TO 14
POINTS 7,8 AND 9 ARE THE SAME,12 AND 13 ARE SAME,
6,10,11 AND 14 ARE SINGULAR

NUMBER OF RELAXATIONS 1

DO YOU REQUIRE A PRINT OUT?IF YES,TYPE 1;NO, TYPE 2
? 2
NUMBER OF RELAXATIONS 2

DO YOU REQUIRE A PRINT OUT?IF YES,TYPE 1;NO, TYPE 2
? 2
NUMBER OF RELAXATIONS 3

DO YOU REQUIRE A PRINT OUT?IF YES,TYPE 1;NO, TYPE 2
? 1
POINT NUMBER  TEMPERATURE

 0              25
 1              25
 2              25
 3              25
 4              25
 5              25
 6              41.6992188
 7              56.6040039
 8              70.9503174
 9              78.7498474
 10             82.2177887
 11             92.6757813
 12             132.515717
 13             145.659256
 14             149.634075
 15             225
 16             225
 17             225
DO YOU WISH ANOTHER RELAXATION?IF YES,TYPE 1;NO,TYPE 2
? 1

NUMBER OF RELAXATIONS 7
POINT NUMBER  TEMPERATURE

 0              25
 1              25
 2              25
 3              25
 4              25
 5              25
 6              42.3116341
 7              60.27947
 8              76.5083813
 9              84.2877596
 10             86.5249977
 11             98.5960515
 12             137.389916
 13             149.800537

NUMBER OF RELAXATIONS 13
POINT NUMBER  TEMPERATURE

 0              25
 1              25
 2              25
 3              25
 4              25
 5              25
 6              43.1049938
 7              61.2390122
 8              77.2762424
 9              84.86553
 10             86.9593152
 11             99.6279694
 12             138.035784
 13             150.251879
```

```
14          152.781518          14          153.115768
15          225                 15          225
16          225                 16          225
17          225                 17          225
HEAT TRANSFER RATE TO CHIMNEY= 1423.112W/M

HEAT TRANSFER RATE FROM CHIMNEY= 1421.75114W/M

STOP AT 460
```

Program nomenclature

QIN inside heat transfer rate
QOT outside heat transfer rate

Program notes

(1) Lines 30 to 60 allocate and store the boundary values and initial estimates contained in the data (line 470).
(2) Line 100 is the iteration counter.
(3) Lines 110 to 240 contain the equations required for solution at various points in the cross-section.
(4) Lines 260 to 380 allow choice of print out of any iteration and escape (when satisfied) to evaluate heat transfer rates in lines 400 and 410.
(5) The results show the third, seventh and thirteenth iterations, after which the heat transfer rate is calculated (the change between the twelfth and thirteenth temperature distributions is small). The closeness of the heat transfer rates is a confirmation of the success of the relaxation process.

Example 7.3 Lumped capacity system

Although the general heat conduction equation is not solved in this work for transient problems in which temperature varies with time, there is one type of problem which can be solved by simple energy balance. If the thermal conductivity of a body is high compared with the heat transfer coefficient at the body surface then the conduction heat transfer within the body is so rapid, compared to the convection heat transfer at the surface, that the temperature distribution within the body may be considered uniform. Such a body is termed a lumped capacity system. A simple example would be a hot metal body being quenched in a cool fluid during heat treatment.

Write a program to determine the rate of change of temperature with time in a lumped capacity system.

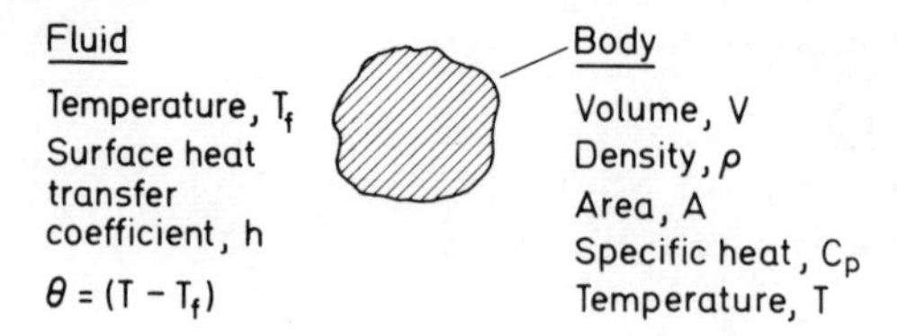

Figure 7.13 Lumped capacity system

Referring to Figure 7.13, in time $d\tau$ the body cools by dT by surface convection, thus

$$-\rho V C_p \, dT = hA(T - T_f)d\tau$$

put $T - T_f = \theta$, then $dT = d\theta$

$$-\rho V C_p \, d\theta = hA\,\theta\,d\tau$$

Consider the time to change in small increments $\Delta\tau$. While this time elapses the temperature falls by $\Delta\theta$

$$\Delta\theta = -\left[\frac{hA\,\theta}{\rho V C_p}\right]\Delta\tau$$

Thus after the time interval $\Delta\tau$, $\theta = \theta + \Delta\theta$ ($\Delta\theta$ is negative) and the incremental method may be repeated.

The exact solution to the problem is

$$\ln\left[\frac{\theta_2}{\theta_1}\right] = -\frac{hA\tau}{mC_p} \tag{7.28}$$

where $m = \rho V$ (the mass of the body).

Test the program to compare the computer solution with the exact solution to the equation. Values of h, A, ρ, V, C_p are given in the program. $\Delta\tau$ may be varied to see how its size affects the results.

```
LIST
 10  PRINT "EXAMPLE 7.3"
 20  PRINT
 30  REM CONSIDER 5 CM ALUMINIUM SPHERES COOLING IN AIR AT 16 C
 35  REM FROM 290 C FOR 2 HOURS WITH CP,RO GIVEN
 40  CP=850
 50  RO=2700
 60  PRINT "WHAT IS THE SURFACE HEAT TRANSFER COEFFICIENT?W?M↑2 K"
 70  INPUT H
 80  V=0.0005236
 90  A=0.0314
 100  TF=16
 110  T1=290
 120  B=0
 130  EL=0
 140  T=T1
 150  PRINT "WHAT TIME INCREMENT?I MINUTES,HOW MANY INCREMENTS?C"
 160  REM C=TOTAL TIME DIVIDED BY THE INCREMENT
 170  INPUT I,C
```

```
180  PRINT "INCREMENT   ELAPSED TIME       TEMPERATURE   "
185  PRINT "                  MIN                C"
190  PRINT
200  PRINT B,EL,T
210  N=C
220  FOR N=1 TO 12
230   B=B+1
240   EL=B*I
250   TH=(T1-TF)
260   DTH=I*((H*A*TH)/(RO*V*CP))*60
270   TH=TH-DTH
280   T=TH+TF
290   PRINT B,EL,T
300   T1=T
310  NEXT N
320  END
```

```
RUN
EXAMPLE 7.3

WHAT IS THE SURFACE HEAT TRANSFER COEFFICIENT?W?M^2 K
? 6
WHAT TIME INCREMENT?I MINUTES,HOW MANY INCREMENTS?C
? 10? 12
INCREMENT   ELAPSED TIME       TEMPERATURE
                MIN                 C


 0              0               290
 1              10              264.224899
 2              20              240.874453
 3              30              219.720578
 4              40              200.556642
 5              50              183.195451
 6              60              167.467423
 7              70              153.218926
 8              80              140.310781
 9              90              128.6169
 10             100             118.023062
 11             110             108.425782
 12             120             99.7313148

WHAT IS THE SURFACE HEAT TRANSFER COEFFICIENT?W?M^2 K
? 6
WHAT TIME INCREMENT?I MINUTES,HOW MANY INCREMENTS?C
? 5? 12
INCREMENT   ELAPSED TIME       TEMPERATURE
                MIN                 C


 0              0               290
 1              5               277.112449
 2              10              264.831063
 3              15              253.127329
 4              20              241.974079
 5              25              231.345421
 6              30              221.216680
 7              35              211.564343
 8              40              202.366
 9              45              193.6003
 10             50              185.246899
 11             55              177.286395
 12             60              169.700313

WHAT IS THE SURFACE HEAT TRANSFER COEFFICIENT?W?M^2 K
? 12
WHAT TIME INCREMENT?I MINUTES,HOW MANY INCREMENTS?C
? 10? 12
```

```
INCREMENT   ELAPSED TIME        TEMPERATURE
                MIN                  C

0                0              290
1                10             238.449797
2                20             196.598220
3                30             162.620574
4                40             135.035463
5                50             112.6402
6                60             94.4583678
7                70             79.6972554
8                80             67.71329
9                90             57.9839815
10               100            50.0851393
11               110            43.67238
12               120            38.4661149

STOP AT 320
```

Program nomenclature

CP specific heat
RO density
V volume
A surface area
TF fluid temperature
T1 initial sphere temperature
EL elapsed time
T instantaneous temperature
TH temperature difference
DTH increment in temperature difference

Program notes

(1) The program is for the material shown in line 30. If different problems were to be solved then lines 40 to 110 could be altered to be on an input basis. In this case C_p = 850 J/kg K, ρ = 2700 kg/m^3, V and A are in m^2 for a 5 cm sphere and the input surface heat transfer coefficient is in W/m^2 K.
(2) Line 120 is an iteration counter.
(3) The choice of increment is important; too long an increment in a cooling process will give very poor results and it would be best to estimate the total cooling time before choosing increments. If this is not possible a range of increments could be tried before deciding on the best value to give the answer to the problem.
(4) The problem is solved in lines 220 to 310.
(5) The results are shown with a time increment of 10 minutes and it can be seen that after one hour the temperature is 167.57°C. A second table with half the time increment shows that after one hour the temperature is 169.7°C. The exact solution after one hour is 171.8°C. The third table doubles the surface heat transfer coefficient so that the

cooling is more rapid and after one hour the temperature is 94.46°C. The exact solution is 104.6°C. The difference has increased and the time interval would have to be reduced considerably to obtain a better result. Further tests could be made with various time intervals and surface heat transfer coefficients.

Example 7.4 Cross flow heat exchanger

In an experiment to determine the forced convection heat transfer relationship for a cylinder in a cross-flow of air the results below were obtained. The experiment consists of blowing cold air at various flow rates over a cylinder repeatedly heated to the same temperature and then allowed to cool to another fixed temperature in a measured time. The cylinder is small enough to be treated as a lumped capacity system (example 7.3) so that

$$\rho V C_p \, \mathrm{d}\theta = -hA\,\theta\,\mathrm{d}\tau$$

hence

$$\ln\left[\frac{\theta_1}{\theta_2}\right] = \frac{hA\tau}{mC_p}$$

where $\theta_1 = (T_{max} - T_{air})$ and $\theta_2 = (T_{min} - T_{air})$

h = surface heat transfer coefficient
A = surface area of cylinder
m = mass of cylinder
C_p = specific heat of cylinder
τ = time to cool from θ_1 to θ_2

The air flow rate is determined by pitot static tube (problem 5.1). The program will first find h and then the Nusselt number, $\mathrm{Nu} = hd/k$ followed by the velocity and the Reynolds number, $\mathrm{Re} = \rho V d/\mu$. The properties of air at the bulk temperature are found in the data block supplied. Evaluate the program with the results:

Time to cool *S*	*Pitot static head* *mm water*
127	40.5
128	33.3
143	23.5
162	14.2
190	7.4
218	4.3
270	1.7

Data

Cylinder maximum temperature	63°C
Cylinder minimum temperature	19°C
Air temperature	14°C
Mass of cylinder	0.1067 kg
Surface area of cylinder	0.00405 m^2
Cylinder diameter	0.01246 m
Specific heat of cylinder	380 J/kg K

For air at 14°C

μ = 0.0000177 kg/ms

ρ = 1.23 kg/m^3

k = 0.0248 W/m K

The necessary reading for this example is in section 7.8.

```
LIST
 10  PRINT "EXAMPLE 7.4"
 20  PRINT
 30  DIM T[7],PS[7],H[7],NU[7],V[7],RE[7]
 40  FOR N=0 TO 6
 50   READ T[N],PS[N]
 60   T1=63
 70   T2=19
 80   TA=14
 90   M=0.1067
 100   A=0.00405
 110   D=0.01246
 120   CP=380
 130   MU=0.0000177
 140   RO=1.23
 150   K=0.0248
 160   REM BY CHOOSING THE SAME MAXIMUM AND MINIMUM TEMPERATURES
 165   REM THE LOG RATIO IS CONSTANT
 170   LR=LOG[(T1-TA)/(T2-TA)]
 180   X=(M*CP)/A
 190   H[N]=LR*X/T[N]
 200   NU[N]=H[N]*D/K
 210   REM CHANGE MM WATER TO M AIR FOR VELOCITY
 220   V[N]=((2*9.81*PS[N])/RO)↑0.5
 230   RE[N]=(RO*V[N]*D)/MU
 240  NEXT N
 250  PRINT "HEAT TRANSFER      NUSSELT          REYNOLDS"
 255  PRINT "COEFFICIENT         NUMBER           NUMBER"
 260  PRINT "  W/M↑2 K"
 270  FOR N=0 TO 6
 280   PRINT H[N],NU[N],RE[N]
 290  NEXT N
 300  PRINT
 310  PRINT "LOG PLOT RE VS NU TO OBTAIN NU = K*RE↑A"
 320  END
 330  DATA 127,40.5
 340  DATA 128,33.3
 350  DATA 143,23.5
 360  DATA 162,14.2
 370  DATA 190,7.4
 380  DATA 218,4.3
 390  DATA 270,1.7
```

```
RUN
EXAMPLE 7.4

HEAT TRANSFER      NUSSELT          REYNOLDS
COEFFICIENT        NUMBER           NUMBER
  W/M↑2 K
 179.919269        9Ø.3949232       22ØØ7.6842
 178.51365Ø        89.6887129       19955.7911
 159.788442        8Ø.28Ø8Ø6Ø       16764.12
 141.Ø47822        7Ø.8651559       13Ø31.3979
 12Ø.261827        6Ø.4218698       94Ø7.25Ø16
 1Ø4.815354        52.6612[illegible]26     7171.Ø221Ø
 84.6286934        42.519Ø935       45Ø8.9Ø717

LOG PLOT RE VS NU TO OBTAIN NU = K*RE↑A

STOP AT 32Ø
```

Program nomenclature

T1	initial temperature
T2	final temperature
TA	air temperature
M	mass of cylinder
A	surface area of cylinder
D	cylinder diameter
CP	specific heat
MU	viscosity
RO	density
K	thermal conductivity
T(N)	time for cooling
H(N)	surface heat transfer coefficient
NU	Nusselt number
PS (N)	pitot-static head
V(N)	air velocity
RE	Reynolds number

Program notes

(1) Lines 40 to 240 evaluate Nu and Re values for each set of readings. The data in lines 60 to 160 is constant and the only measured variables are the pitot-static head and the cooling time, given as data in lines 330 to 390 and read in at line 50.

(2) By measuring the cooling time between two fixed temperatures the ratio $\ln(\theta_1/\theta_2)$ is constant and the calculation of Nu (lines 170 to 200) and Re (lines 220 and 230) is made very simple.

(3) The log plot shows that the relation in this heat exchanger is

$$\mathrm{Nu} = 0.6\mathrm{Re}^{0.5}$$

Compare this with standard results.

Example 7.5 Simple heat exchanger

Water flows through tubes in a heat exchanger in which 50 kW are transferred by the walls to the water. The tube walls may be considered to be at a constant temperature of 25°C and the water enters at 10°C and leaves at 20°C. If the tube diameter is varied the surface heat transfer coefficient will change thus altering the length of tube required.

Write a program to determine some of the options available in single and multi-tube designs. In all designs the water flows in a single direction through the heat exchanger and the flow rate of water is divided evenly between the tubes in multi-tube options.

The required heat transfer rate

$$\dot{Q} = [\dot{m}\, C_p\, (T_{\text{out}} - T_{\text{in}})]_{\text{water}}$$

If the flow is turbulent (Re > 2300), then

$$\text{Nu} = 0.023\ \text{Re}^{0.8}\ \text{Pr}^{0.4}$$

hence

$$h = \frac{k}{d}\ 0.023\ \text{Re}^{0.8}\ \text{Pr}^{0.4}$$

$$\theta_{\text{mean}} = \frac{\theta_1 - \theta_2}{\ln[\theta_1/\theta_2]}$$

For a single tube

$$\dot{Q} = h A\, \theta_{\text{mean}} \quad \text{with } A = \pi\, dL$$

For a multi-tube case

$$\dot{Q} \text{ per tube} = \dot{Q}/\text{number of tubes}$$

and

$$\dot{m} \text{ per tube} = \dot{m}/\text{number of tubes}$$

The necessary reading for this example is in section 7.9.

```
LIST
 10  PRINT "EXAMPLE 7.5"
 20  PRINT
 30  TW=25
 40  T1=10
 50  T2=20
 60  Q=50
 70  PRINT "WATER PROPERTIES AT BULK TEMP 15 C FROM DATA BLOCK"
 80  K=0.595
 90  MU=0.001152
 100  CP=4.188
```

```
110  PR=(1000*MU*CP)/K
120  PRINT "SINGLE TUBE OPTIONS,DIAMETER D=1 TO 5 CM"
130  PRINT
140  PRINT "DIAMETER           LENGTH         REYNOLDS NUMBER"
145  PRINT "  CM                  M"
150  PRINT
160  MW=Q/(CP*(T2-T1))
170  REM REYNOLDS NUMBER RO*V*D/MU=4*M/3.142*D*MU
180  FOR D=1 TO 5
190   RE=(4*MW)/(3.142*(D/100)*MU)
200   LTD=((25-10)-(25-20))/LOG[(25-10)/(25-20)]
210   H=(0.023*(PR↑0.4)*K*(RE↑0.8))/D
220   L=Q*1000/(H*3.142*D*LTD)
230   PRINT D,L,RE
240  NEXT D
250  PRINT "NEGLECT ANY OPTIONS WITH RE<2300 AS EQUATIONS INVALID"
260  PRINT
270  PRINT "MULTITUBE OPTIONS,NUMBER N=6 TO 10WITH D=1 TO 3 CM"
280  FOR D=1 TO 3
290   PRINT
300   PRINT "DIAMETER       NUMBER              LENGTH      REYNOLDS NUMBER"
305   PRINT "  CM                                  M"
310   PRINT
320   FOR N=6 TO 10
330    MW=Q/(CP*(T2-T1)*N)
340    RE=(4*MW)/(3.142*(D/100)*MU)
350    LTD=((25-10)-(25-20))/LOG[(25-10)/(25-20)]
360    H=(0.023*(PR↑0.4)*K*(RE↑0.8))/D
370    L=(1000*Q)/(H*3.142*D*LTD*N)
380    PRINT D,N,L,RE
390   NEXT N
400   PRINT "NEGLECT ANY OPTIONS WITH RE<2300 AS EQUATIONS INVALID"
410  NEXT D
420  END

RUN
EXAMPLE 7.5

WATER PROPERTIES AT BULK TEMP 15 C FROM DATA BLOCK
SINGLE TUBE OPTIONS,DIAMETER D=1 TO 5 CM

DIAMETER          LENGTH         REYNOLDS NUMBER
  CM                M

 1               4.43092980       131936.410
 2               7.71469687       65968.2050
 3               10.6706745       43978.8033
 4               13.4320674       32984.1025
 5               16.0572391       26387.2820
NEGLECT ANY OPTIONS WITH RE<2300 AS EQUATIONS INVALID

MULTITUBE OPTIONS,NUMBER N=6 TO 10WITH D=1 TO 3 CM

DIAMETER       NUMBER              LENGTH      REYNOLDS NUMBER
  CM                                 M

 1             6                 3.09645391      21989.4017
 1             7                 3.00244639      18848.0586
 1             8                 2.92332346      16492.0513
 1             9                 2.85526464      14659.6011
 1             10                2.79572770      13193.6410
NEGLECT ANY OPTIONS WITH RE<2300 AS EQUATIONS INVALID

DIAMETER       NUMBER              LENGTH      REYNOLDS NUMBER
  CM                                 M

 2             6                 5.39123938      10994.7
 2             7                 5.22756279      9424.02929
 2             8                 5.08980177      8246.02563
 2             9                 4.97130449      7329.80056
 2             10                4.86764465      6596.82050
NEGLECT ANY OPTIONS WITH RE<2300 AS EQUATIONS INVALID
```

```
DIAMETER     NUMBER          LENGTH        REYNOLDS NUMBER
  CM                           M

3             6             7.45695673       7329.80056
3             7             7.23056551       6282.68619
3             8             7.04001973       5497.35042
3             9             6.87611880       4886.53371
3             10            6.73274046       4397.88033
NEGLECT ANY OPTIONS WITH RE<2300 AS EQUATIONS INVALID

STOP AT 420
```

Program nomenclature

TW	wall temperature
T1	water inlet temperature
T2	water outlet temperature
Q	heat transfer rate
K	thermal conductivity
MU	viscosity
CP	specific heat
PR	Prandtl number
MW	mass flow rate of water
D	tube diameter
RE	Reynolds number
LTD	log mean temperature difference
H	surface heat transfer coefficient
L	tube length

Program notes

(1) The program aims to show the effect on heat exchanger length of varying the tube size and of using several tubes instead of a single tube. The heat transfer rate for all the 'designs' is the same, as are the water temperature rise and the wall temperature. It might be a range of designs for a condenser.
(2) The problem is solved for a single tube in lines 160 to 220 evaluating the water flow rate, the Reynolds number, the heat transfer coefficient and the log mean temperature difference. Similarly (in lines 280 to 370) the problem is solved for multi-tube options. The difference between these two programs is the division of the water flow rate between the N tubes so that each tube effects $\dot{Q}/N$ of heat transfer.
(3) The results in the single tube option show that the small tubes give high Reynolds numbers, hence shorter lengths, but not in direct proportion to diameter. Halving the diameter does not achieve half the length. In the multi-tube options the same effect can be seen by comparing like lines in the three cases. It can also be seen that increasing the number of tubes has less effect on length than changing

the diameter. For example, one tube, 1 cm in diameter, needs 4.43 m but 10 tubes, 1 cm in diameter, still need 2.8 m in length (i.e. 28 m in total). Obviously, considerable thought is necessary before choosing a particular heat exchanger design.

Example 7.6 Domestic radiator performance

A radiator placed centrally in a large room has an emissivity ϵ. Heat transfer to the surroundings is by radiation and free convection from the two vertical faces of height L at temperature T_1. The surroundings are at temperature T_2. Write a program to determine the total heat transfer rate per unit width of radiator. Evaluate the program with

$$\epsilon = 0.7, L = 0.8, T_1 = 80°C, T_2 = 20°C$$

For radiation $\sigma = 5.67 \times 10^{-8}$ W/m^2 K^4 and the radiator may be treated as a grey body in large surroundings.

For free convection use the relations given in section 7.8, which give average values over the whole surface.

This program could be extended to determine the area of heating surface required to heat a house when the losses have been determined by the methods of example 7.1 and problem 7.1.

The necessary reading is in sections 7.3, 7.8 and 7.13.

```
LIST
 10  PRINT "EXAMPLE 7.6"
 20  PRINT
 30  REM FIND GRASHOF NUMBER TO DETERMINE FLOW REGIME
 40  PRINT "FLUID PROPERTIES AT T FILM 50 C FROM DATA BLOCK"
 50  DIM T[10],VS[10],CP[10],K[10],MU[10],PR[10]
 60  FOR N=0 TO 9
 70   READ T[N],VS[N],CP[N],K[N],MU[N],PR[N]
 80  NEXT N
 90  C=0:: D=0:: F=0:: G=0
 100  PRINT "TEMPERATURE?"
 110  INPUT A
 120  FOR J=0 TO 5
 130   X=1
 140   FOR N=0 TO 5
 150    IF N=J THEN GOTO 170
 160    X=X*(A-T[N])/(T[J]-T[N])
 170   NEXT N
 180   C=C+X*VS[J]
 190   D=D+X*CP[J]
 200   F=F+X*MU[J]
 210   G=G+X*K[J]
 220  NEXT J
 230  RO=1/C
 240  CP=D
 250  MU=F/1000
 260  K=G/1000
 270  B=1/(50+273)
 280  L=0.8
 290  T1=80
 300  T2=20
 310  GR=(B*9.81*(T1-T2)*(RO↑2)*(L↑3)*CP)/(MU*K)
 320  PRINT "GRASHOF NUMBER=";GR
 330  IF GR<1000000000 THEN GOTO 360
```

```
340  NU=0.13*(GR↑0.33)
350  GOTO 370
360  NU=0.59*(GR↑0.25)
370  H=(NU*K)/L
380  QC=H*2*L*1*(T1-T2)
390  PRINT
400  PRINT "CONVECTIVE HEAT TRANSFER=":QC;"KW/M"
410  PRINT
420  PRINT "RADIATION FROM A SMALL GREY BODY IN LARGE SURROUNDINGS"
430  E=0.7
440  S=5.67E-11
450  QR=2*L*E*S*((353↑4)-(293↑4))
460  PRINT "RADIATION HEAT TRANSFER=";QR;"KW/M"
470  PRINT
480  Q=QC+QR
490  PRINT "HEAT TRANSFER RATE PER UNIT WIDTH=";Q;"KW"
500  END


2310  DATA -100,0.488,1.01,0.016,0.012,0.75
2320  DATA 0,0.773,1.01,0.024,0.017,0.72
2330  DATA 100,1.057,1.02.0.032,0.022,0.7
2340  DATA 200,1.341,1.03,0.039,0.026,0.69
2350  DATA 300,1.624,1.05,0.045,0.03,0.69
2360  DATA 400,1.908,1.07,0.051,0.033,0.7
2370  DATA 500,2.191,1.1,0.056,0.036,0.7
2380  DATA 600,2.473,1.12,0.061,0.039,0.71
2390  DATA 700,2.756,1.14,0.066,0.042,0.72
2400  DATA 800,3.039,1.16,0.071,0.044,0.73
```

```
RUN
EXAMPLE 7.6

FLUID PROPERTIES AT T FILM 50 C FROM DATA BLOCK
TEMPERATURE?
? 50
GRASHOF NUMBER= 2049728970

CONVECTIVE HEAT TRANSFER= 0.518171292KW/M

RADIATION FROM A SMALL GREY BODY IN LARGE SURROUNDINGS
RADIATION HEAT TRANSFER= 0.518024487KW/M

HEAT TRANSFER RATE PER UNIT WIDTH= 1.03619578KW

STOP AT 500
```

Program nomenclature

RO density
CP specific heat
MU viscosity
K thermal conductivity
B coefficient of cubical expansion
L height of radiator
T1 surface temperature
T2 air temperature
GR Grashof number
NU Nusselt number
H surface heat transfer coefficient
QC convective heat transfer

E emissivity
S Stefan–Boltzmann constant
QR radiation heat transfer
Q total heat transfer

Program notes

(1) The data for this problem is determined by an interpolation routine in lines 50 to 260. An easier alternative would be to put the data in directly, or preferably to use a keyboard input routine. In many cases these pieces of data would probably be constant so that the heating area could be determined from a calculation of the heat losses in any particular situation (see example 7.1 and problem 7.1).
(2) The Grashof number test is made to see if flow over the radiator surface is laminar or turbulent and it is seen that it just becomes turbulent. The equations used for Nusselt numbers give average values and allow for any flow regime that occurs (lines 300 to 350).
(3) The radiator is treated as a grey body in large surroundings but this approach could be considered presumptive if the radiator is against a wall, since there is bound to be reflection from the wall which is re-incident on the radiator. The solution of this problem would be more involved and the answer obtained would not be greatly different since any radiation re-incident on the radiator would be redistributed to the whole room, which completely surrounds the radiator.
(4) Try varying the radiator surface temperature to determine the effect on each component of heat transfer and the total heat transfer. Perhaps consider the temperature decrease in the radiator water flow as the energy is transferred to the room. What is the flow rate of water per kW of energy transfer with a reasonable temperature drop? How many radiators could you have in series before the temperature is too low to be effective as a heater? Can you control by temperature, flow rate or an on/off situation? What about off-peak electric storage heaters; how does their energy output fall in a day off charge?

Example 7.7 Radiation shields

A molten metal surface 5 m by 2 m at 1500°C radiates heat to another surface of equal size at 50°C which is situated 5 m above the metal surface and parallel to the metal surface.

To reduce the heat transfer from the molten metal to the upper surface a radiation shield is placed between the two surfaces. The shield is of equal area to the surfaces and placed parallel to the surfaces. Write a program to determine the optimum position of the shield for minimum heat transfer. The emissivity of the molten metal surface is

0.8, of the shield is 0.5 and of the top surface is 0.6. Having decided the optimum position determine the effect of varying the emissivity of the shield when placed centrally. Figure 7.10 shows the geometric factors involved.

Using this Hottel chart the following values of geometric factors are obtained.

Shield position from hot surface (m)	1	2	3	4	2.5
Geometric factor from hot surface to shield	0.535	0.26	0.190	0.130	0.23
Geometric factor from shield to cool surface	0.130	0.190	0.26	0.535	0.23

With no shield the heat transfer rate is 500 kW.

The problem is a six-resistance situation (Figure 7.14).

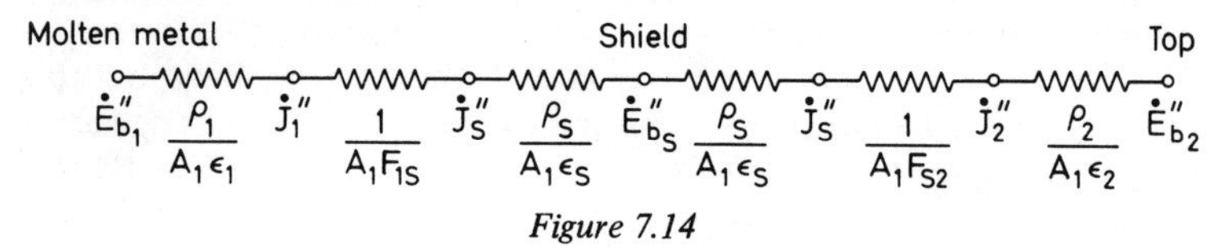

Figure 7.14

The heat transfer from molten metal to shield is given by

$$\dot{Q}_{1s} = \frac{\sigma(T_1{}^4 - T_s{}^4)}{\dfrac{\rho_1}{A_1\epsilon_1} + \dfrac{1}{A_1F_{1s}} + \dfrac{\rho_s}{A_1\epsilon_s}}$$

$$= \frac{\sigma(T_1{}^4 - T_s{}^4)}{R_1}$$

The heat transfer from shield to top surface is given by

$$\dot{Q}_{s2} = \frac{\sigma(T_s{}^4 - T_2{}^4)}{\dfrac{\rho_s}{A_1\epsilon_s} + \dfrac{1}{A_1F_{s2}} + \dfrac{\rho_2}{A_1\epsilon_2}}$$

$$= \frac{\sigma(T_s{}^4 - T_2{}^4)}{R_2}$$

By equating these heat transfers

$$T_s{}^4 = \frac{R_2T_1^4 + R_1T_2{}^4}{R_1 + R_2}$$

which can be solved for T_s. T_s is then substituted back to determine the heat transfer rate.

The necessary reading for this example is in section 7.14.

```
LIST
 1Ø  PRINT "EXAMPLE 7.7"
 2Ø  PRINT
 3Ø  DIM A[4],ES[4],TS[4],Q[4]
 4Ø  E1=Ø.8
 5Ø  E2=Ø.6
 6Ø  T1=1773
 7Ø  T2=323
 8Ø  N=Ø
 9Ø  PRINT "EMISSIVITY OF SHIELD?"
 1ØØ  INPUT ES[N]
 11Ø  PRINT "SHIELD DISTANCE ABOVE HOT SURFACE?"
 12Ø  INPUT A[N]
 13Ø  PRINT "GEOMETRIC FACTOR 1 TO S?"
 14Ø  REM USE HOTTEL CHART
 15Ø  INPUT F1
 16Ø  PRINT "GEOMETRIC FACTOR S TO 2?"
 17Ø  REM USE HOTTEL CHART
 18Ø  INPUT F2
 19Ø  REM HEAT TRANSFER TO SHIELD Q1=(T1↑4-TS↑4)/SUM OF RESISTANCES
 2ØØ  REM HEAT TRANSFER FROM SHIELD Q2=(TS↑4-T2↑4)/SUM OF RESISTANCES
 21Ø  PRINT " EQUATE HEAT TRANSFERS TO AND FROM SHIELD TO GET TS(N)"
 22Ø  R1=((1-E1)/E1+1/F1+(1-ES[N])/ES[N])
 23Ø  R2=((1-ES[N])/ES[N]+1/F2+(1-E2)/E2)
 24Ø  TS[N]=((R2*(T1↑4)+R1*(T2↑4))/(R1+R2))↑Ø.25
 25Ø  Q[N]=5*2*5.67E-11*((T1↑4)-(TS[N]↑4))/R1
 26Ø  REM RUN AT POSITIONS A=1,2,3 AND 4 WITH ES(N)=Ø.5,THEN ON
 265  REM SECOND RUN VARY SHIELD EMISSIVITY
 27Ø  PRINT "DO YOU WANT ANOTHER RUN?IF YES,TYPE 1;IF NO,2"
 28Ø  INPUT X
 29Ø  N=N+1
 3ØØ  IF X=1 THEN GOTO 9Ø
 31Ø  PRINT "SHIELD          SHIELD          SHIELD          HEAT"
 32Ø  PRINT "POSITION        EMISSIVITY      TEMPERATURE     TRANSFER"
 325  PRINT "M ABOVE HOT                     K               KW"
 33Ø  FOR N=Ø TO 3
 34Ø   PRINT A[N],ES[N],TS[N],Q[N]
 35Ø  NEXT N
 36Ø  PRINT "ESTIMATE THE OPTIMUM SHIELD POSITION FROM THESE RESULTS"
 365  PRINT "THEN A=2.5 FOR SECOND RUN"
 37Ø  PRINT "NOW VARY EMISSIVITY OF SHIELD,ES(N)=Ø.9,Ø.7,Ø.3 AND Ø.1"
 38Ø  PRINT " IF THE EMISSIVITY VARIATION IS COMPLETED TYPE 1,IF NOT,TYPE 2"
 39Ø  INPUT Y
 4ØØ  IF Y=2 THEN GOTO 8Ø
 41Ø  END


RUN
EXAMPLE 7.7

EMISSIVITY OF SHIELD?
? Ø.5
SHIELD DISTANCE ABOVE HOT SURFACE?
? 1
GEOMETRIC FACTOR 1 TO S?
? Ø.535
GEOMETRIC FACTOR S TO 2?
? Ø.13
 EQUATE HEAT TRANSFERS TO AND FROM SHIELD TO GET TS(N)
DO YOU WANT ANOTHER RUN?IF YES,TYPE 1;IF NO,2
? 1
EMISSIVITY OF SHIELD?
? Ø.5
SHIELD DISTANCE ABOVE HOT SURFACE?
? 2
```

```
GEOMETRIC FACTOR 1 TO S?
? Ø.26
GEOMETRIC FACTOR S TO 2?
? Ø.19
 EQUATE HEAT TRANSFERS TO AND FROM SHIELD TO GET TS(N)
DO YOU WANT ANOTHER RUN?IF YES,TYPE 1;IF NO,2
? 1
EMISSIVITY OF SHIELD?
? Ø.5
SHIELD DISTANCE ABOVE HOT SURFACE?
? 3
GEOMETRIC FACTOR 1 TO S?
? Ø.19
GEOMETRIC FACTOR S TO 2?
? Ø.26
 EQUATE HEAT TRANSFERS TO AND FROM SHIELD TO GET TS(N)
DO YOU WANT ANOTHER RUN?IF YES,TYPE 1;IF NO,2
? 1
EMISSIVITY OF SHIELD?
? Ø.5
SHIELD DISTANCE ABOVE HOT SURFACE?
? 4
GEOMETRIC FACTOR 1 TO S?
? Ø.13
GEOMETRIC FACTOR S TO 2?
? Ø.535
 EQUATE HEAT TRANSFERS TO AND FROM SHIELD TO GET TS(N)
DO YOU WANT ANOTHER RUN?IF YES,TYPE 1;IF NO,2
? 2
SHIELD          SHIELD          SHIELD          HEAT
POSITION        EMISSIVITY      TEMPERATURE     TRANSFER
M ABOVE HOT                          K             KW
 1               Ø.5             165Ø.13Ø32      448.5282
 2               Ø.5             1545.Ø6624      465.392Ø42
 3               Ø.5             1459.36287      465.392Ø42
 4               Ø.5             1294.48216      448.5282
ESTIMATE THE OPTIMUM SHIELD POSITION FROM THESE RESULTS
THEN A=2.5 FOR SECOND RUN
NOW VARY EMISSIVITY OF SHIELD,ES(N)=Ø.9,Ø.7,Ø.3 AND Ø.1
 IF THE EMISSIVITY VARIATION IS COMPLETED TYPE 1,IF NOT,TYPE 2
? 2
EMISSIVITY OF SHIELD?
? Ø.9
SHIELD DISTANCE ABOVE HOT SURFACE?
? 2.5
GEOMETRIC FACTOR 1 TO S?
? Ø.23
GEOMETRIC FACTOR S TO 2?
? Ø.23
 EQUATE HEAT TRANSFERS TO AND FROM SHIELD TO GET TS(N)
DO YOU WANT ANOTHER RUN?IF YES,TYPE 1;IF NO,2
? 1
EMISSIVITY OF SHIELD?
? Ø.7
SHIELD DISTANCE ABOVE HOT SURFACE?
? 2.5
GEOMETRIC FACTOR 1 TO S?
? Ø.23
GEOMETRIC FACTOR S TO 2?
? Ø.23
 EQUATE HEAT TRANSFERS TO AND FROM SHIELD TO GET TS(N)
DO YOU WANT ANOTHER RUN?IF YES,TYPE 1;IF NO,2
? 1
EMISSIVITY OF SHIELD?
? Ø.3
SHIELD DISTANCE ABOVE HOT SURFACE?
? 2.5
GEOMETRIC FACTOR 1 TO S?
? Ø.23
GEOMETRIC FACTOR S TO 2?
? Ø.23
```

```
 EQUATE HEAT TRANSFERS TO AND FROM SHIELD TO GET TS(N)
DO YOU WANT ANOTHER RUN?IF YES,TYPE 1;IF NO,2
? 1
EMISSIVITY OF SHIELD?
? 0.1
SHIELD DISTANCE ABOVE HOT SURFACE?
? 2.5
GEOMETRIC FACTOR 1 TO S?
? 0.23
GEOMETRIC FACTOR S TO 2?
? 0.23
 EQUATE HEAT TRANSFERS TO AND FROM SHIELD TO GET TS(N)
DO YOU WANT ANOTHER RUN?IF YES,TYPE 1;IF NO,2
? 2
SHIELD          SHIELD          SHIELD          HEAT
POSITION        EMISSIVITY      TEMPERATURE     TRANSFER
M ABOVE HOT                          K              KW
 2.5             0.9             1506.837        569.095661
 2.5             0.7             1505.90957      534.582848
 2.5             0.3             1502.05856      391.960245
 2.5             0.1             1496.90188      202.691946
ESTIMATE THE OPTIMUM SHIELD POSITION FROM THESE RESULTS
THEN A=2.5 FOR SECOND RUN
NOW VARY EMISSIVITY OF SHIELD,ES(N)=0.9,0.7,0.3 AND 0.1
 IF THE EMISSIVITY VARIATION IS COMPLETED TYPE 1,IF NOT,TYPE 2
? 1

STOP AT 410
```

Program nomenclature

E1 emissivity of molten metal
E2 emissivity of top surface
T1 temperature of molten metal
T2 temperature of top surface
R1 resistance, metal to shield
R2 resistance, shield to top
TS shield temperature
Q heat transfer rate

Program notes

(1) As in the earlier programs the data in lines 40 to 70 could be supplied on an input basis if the program was to be used for similar problems.
(2) The Hottel chart data has been determined as input data rather than 'programming' the contours of the chart. Contour programs are possible but tedious except for repeated use. Examples of these are the steam and refrigerant data blocks.
(3) Lines 220 and 230 find the two total resistances used to determine shield temperature and heat transfer in lines 240 and 250. The program is then run for the various options.
(4) It is seen that the shield should be placed as near to the metal surface or the top surface as possible to reduce the heat transfer

between these two surfaces. It is also seen that the shield emissivity should be as low as possible.

(5) The results for this problem do not include heat transfers to the other surfaces, which must obviously surround the metal surface, and so the answer is incomplete in that it does not give the total heat loss from the metal surface. To reduce this heat loss, low emissivity should be chosen for all the surroundings so that the majority of the radiation emitted by the metal surface is reflected.

(6) It can be seen that some of the options can increase the heat transfer between these two surfaces.

Example 7.8 Fin heat transfer

Write a program to determine the temperature distribution in a slender rectangular fin using finite difference methods and then compare the heat transfer rate from the fin with that given by analytical solution in section 7.11.

Data

Wall temperature 100°C
Fluid temperature 10°C
Fin length 0.06 m, thickness 0.003 m, unit width
Thermal conductivity of fin material 50 W/m K
Surface heat transfer coefficient 40 W/m² K

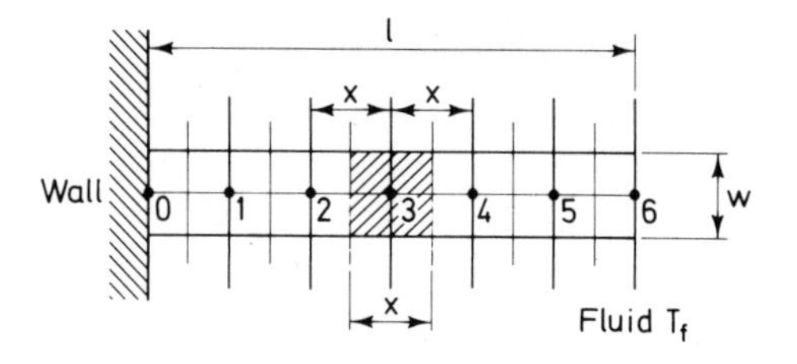

Figure 7.15

By similar methods to those of example 7.2, mesh point equations are written. In this case of one-dimensional conduction the mesh does not need to be rectangular (Figure 7.15). Initial temperatures are allocated and the mesh is solved until satisfactory successive results are achieved. Three equations are required.

At the wall

$$T(0) = T_{\text{wall}}$$

In the centre (for example point 3)

$$\dot{Q}_{23} + \dot{Q}_{43} + \dot{Q}_{f3} = 0$$

$$k(w.1)\left(\frac{T_2 - T_3}{x}\right) + k(w.1)\left(\frac{T_4 - T_3}{x}\right) + 2\,h\,(x.1)(T_f - T_3) = 0$$

giving

$$T(3) = \left[T(2) + T(4) + \left(\frac{2hx^2}{kw}\right)\,T_f\right] \Big/ 2\left[1 + \frac{hx^2}{kw}\right]$$

At the far end (neglecting heat transfer from the end as $\mathrm{d}T/\mathrm{d}x = 0$)

$$k(w.1)\left(\frac{T_5 - T_6}{x}\right) + 2\left[\frac{x}{2}\,.\,1\right]h(T_f - T_6) = 0$$

giving

$$T(6) = \left[T(5) + \left(\frac{hx^2}{kw}\right)T_f\right] \Big/ \left[1 + \frac{hx^2}{kw}\right]$$

The heat transfer rate from the wall to the fin is given by

$$\dot{Q}_{01} = k(w.1)\,\frac{(T(0) - T(1))}{x} \quad \text{per unit width}$$

The necessary reading for this example is in section 7.7.

```
LIST
 10  PRINT "EXAMPLE 7.8"
 20  PRINT
 30  DIM T[7]
 40  FOR N=0 TO 6
 50   READ T[N]
 60  NEXT N
 70  PRINT "T(0)= WALL TEMPERATURE(100 C)"
 80  TF=10
 90  L=0.06
 100  W=0.003
 110  K=50
 120  H=40
 130  PRINT "USE 7 NODES"
 140  PRINT
 150  X=L/6
 160  F=(H*(X↑2))/(K*W)
 170  REM THERE ARE 6 POINTS TO BE RELAXED,N=1 TO 6
 180  REM THE POINTS 1 TO 5 ARE SIMILAR BUT 6 IS SINGULAR
 190  A=0
 200  FOR N=1 TO 5
 210   T[N]=(T[N-1]+T[N+1]+2*F*TF)/(2*(1+F))
 220  NEXT N
 230  N=6
 240  T[N]=(T[5]+F*TF)/(1+F)
 250  A=A+1
 260  PRINT "NUMBER OF ITERATIONS";A
 270  PRINT
 280  PRINT "DO YOU REQUIRE A PRINTOUT?IF YES,TYPE 1;NO,TYPE 2"
 290  INPUT Y
 300  IF Y=2 THEN GOTO 200
```

```
310  PRINT
320  PRINT "POINT NUMBER     TEMPERATURE(C)"
330  FOR N=0 TO 6
340   PRINT N,T[N]
350  NEXT N
360  PRINT "DO YOU REQUIRE ANOTHER ITERATION?IF YES,TYPE 1;NO,TYPE 2"
370  INPUT Z
380  IF Z=1 THEN GOTO 200
390  Q=(K*W*1*(100-T[1]))/X
400  PRINT
410  PRINT "HEAT TRANSFER RATE FROM FIN=";Q;"W"
420  END
430  DATA 100,50,50,50,50,50,50
```

```
RUN
EXAMPLE 7.8

T(0)= WALL TEMPERATURE(100 C)
USE 7 NODES

NUMBER OF ITERATIONS 1

DO YOU REQUIRE A PRINTOUT?IF YES,TYPE 1;NO,TYPE 2
? 2
NUMBER OF ITERATIONS 2

DO YOU REQUIRE A PRINTOUT?IF YES,TYPE 1;NO,TYPE 2
? 2
NUMBER OF ITERATIONS 3

DO YOU REQUIRE A PRINTOUT?IF YES,TYPE 1;NO,TYPE 2
? 1

POINT NUMBER     TEMPERATURE(C)
 0               100
 1               80.4711654
 2               66.9954526
 3               58.1455458
 4               52.4789721
 5               49.2249534
 6               48.2061234
DO YOU REQUIRE ANOTHER ITERATION?IF YES,TYPE 1;NO,TYPE 2
? 1
NUMBER OF ITERATIONS 6

DO YOU REQUIRE A PRINTOUT?IF YES,TYPE 1;NO,TYPE 2
? 1

POINT NUMBER     TEMPERATURE(C)
 0               100
 1               82.6052186
 2               69.5739330
 3               60.1871197
 4               53.8912343
 5               50.3198757
 6               49.2726062
DO YOU REQUIRE ANOTHER ITERATION?IF YES,TYPE 1;NO,TYPE 2
? 1
NUMBER OF ITERATIONS 9

DO YOU REQUIRE A PRINTOUT?IF YES,TYPE 1;NO,TYPE 2
? 1

POINT NUMBER     TEMPERATURE(C)
 0               100
 1               83.1393329
 2               70.3904332
 3               61.1232019
 4               54.8698684
 5               51.3056552
 6               50.2327811
```

```
DO YOU REQUIRE ANOTHER ITERATION?IF YES,TYPE 1;NO,TYPE 2
? 2

HEAT TRANSFER RATE FROM FIN= 252.91W

STOP AT 420
```

Program nomenclature

TF fluid temperature
L fin length
W fin thickness
K thermal conductivity
H surface heat transfer coefficient
Q heat transfer from fin

Program notes

(1) This problem is similar to example 7.2 in which an iteration technique is used to solve simultaneous equations. In lines 190 to 240 the two equations to be solved are programmed and in line 250 an iteration counter is used. The program allows iterations to proceed, with print-out choice, until the user is satisfied that no improvement in result is achieved with further work. The heat transfer rate is then determined.
(2) The results of the third, sixth and ninth iterations are shown together with the heat transfer rate after nine iterations. The analytical result for the heat transfer rate is 275.4 W. If further iterations are made it will be possible to get closer to this result but if the mesh is too large, i.e. there need to be more elements in the model, it would be preferable to change the mesh size.
(3) This technique could be adapted for use for fins of tapering cross-section or of annular form.

Example 7.9 NTU method

A stream of fluid is to be cooled by water flowing in a single tube with the fluid in counter flow in an annular space outside the tube. The heat transfer coefficient on the fluid side is 3 kW/m^{2}K. The flow rate of fluid is 0.5 kg/s with an entry temperature of 130°C and specific heat 2.24 kJ/kg K. The flow rate of water in the tube is 0.3 kg/s with an entry temperature of 10°C. It may be assumed that the mean bulk temperature of the water is 50°C in all cases.

Write a program to determine the fluid exit temperature for tube diameters of 2 cm, 3 cm and 4 cm with a total length of tube 20 m.

This is a problem in which only the fluid and water inlet temperatures are known and may be solved by the NTU method. Initially

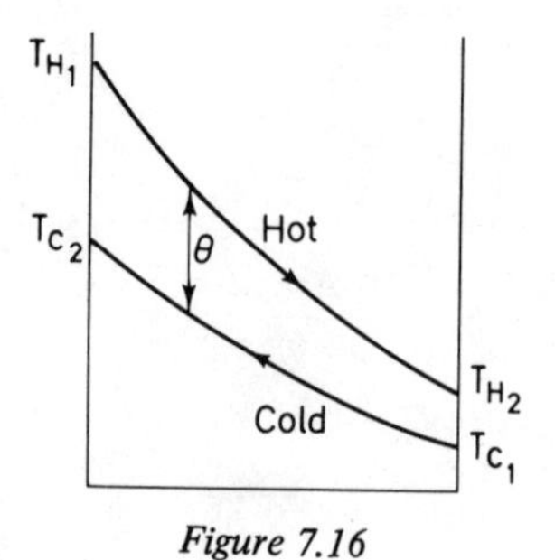

Figure 7.16

the capacity ratio C must be determined which implies deciding which is the greater of $(\dot{m}C_p)_{\text{water}}$ and $(\dot{m}\, C_p)_{\text{fluid}}$. In this case

$$(\dot{m}\, C_p)_{\text{water}} = 0.3 \times 4.182 = 1.2546$$

and

$$(\dot{m}\, C_p)_{\text{fluid}} = 0.5 \times 2.24 = 1.12$$

hence

$$C = \frac{1.12}{1.2546} = 0.8927$$

For this case, in a counter flow heat exchanger (Figure 7.16)

$$E = \frac{t_{H_1} - t_{H_2}}{t_{H_1} - t_{c_2}} = \frac{1 - e^{\text{NTU}(1 - C)}}{1 - Ce^{\text{NTU}(1 - C)}}$$

where

$$\text{NTU} = \frac{UA}{(\dot{m}\, C_p)_{\text{fluid}}}$$

and $A = \pi dL$ (d = tube diameter and L = tube length).
U is found from

$$\frac{1}{U} = \frac{1}{h_{\text{fluid}}} + \frac{1}{h_{\text{water}}}$$

where $h_{\text{water}} = \frac{k}{d}\,(0.023\ \text{Re}^{0.8}\ \text{Pr}^{0.4})$

The necessary reading for this example is in section 7.10.

```
LIST
 10  PRINT "EXAMPLE 7.9"
 20  PRINT
 30  PRINT "WHAT IS THE TUBE DIAMETER IN M?"
 40  INPUT D
 50  PRINT "WHAT IS THE TUBE LENGTH IN M?"
 60  INPUT L
 70  PRINT "WATER PROPERTIES AT 50 C FROM DATA BLOCK"
 80  RO=1/0.00101
 90  MU=0.000544
 100  K=0.000643
 110  PR=3.53
 120  REM REYNOLDS NUMBER=4*M/3.142*D*MU
 130  PRINT "WHAT IS THE CALCULATED CAPACITY RATIO?"
 140  INPUT C
 150  RE=(4*0.3)/(3.142*D*MU)
 160  PRINT "RE=";RE;"IF RE<2300 SOLUTION IS INVALID"
 170  H=(0.023*(RE↑0.8)*(PR↑0.4)*K)/D
 180  PRINT "WATER HEAT TRANSFER COEFFICIENT=";H;"KW/M↑2 K"
 190  PRINT
 200  U=1/((1/H)+(1/3))
 210  A=3.142*D*L
 220  NTU=(U*A)/1.12
 230  PRINT "NTU=";NTU
 240  E1=1-(1/EXP[NTU*(1-C)])
 250  E2=1-(C/EXP[NTU*(1-C)])
 260  E=E1/E2
 270  PRINT "EFFECTIVENESS";E
 280  T2=130-(120*E)
 290  PRINT
 300  D=D*100
 310  PRINT "DIAMETER OF TUBE=";D;"CM"
 320  PRINT "LENGTH OF TUBE=";L;"M"
 330  PRINT "FLUID OUTLET TEMPERATURE=";T2;" C"
 340  PRINT
 350  PRINT "DO YOU WISH TO TRY ANOTHER DESIGN? IF YES,TYPE 1;IF NO,TYPE 2"
 360  INPUT X
 370  IF X=1 THEN GOTO 30
 380  END
```

```
RUN
EXAMPLE 7.9

WHAT IS THE TUBE DIAMETER IN M?
? .02
WHAT IS THE TUBE LENGTH IN M?
? 20
WATER PROPERTIES AT 50 C FROM DATA BLOCK
WHAT IS THE CALCULATED CAPACITY RATIO?
? .8927
RE= 35103.1565IF RE<2300 SOLUTION IS INVALID
WATER HEAT TRANSFER COEFFICIENT= 5.30024117KW/M↑2 K

NTU= 2.14968252
EFFECTIVENESS 0.707415878

DIAMETER OF TUBE= 2CM
LENGTH OF TUBE= 20M
FLUID OUTLET TEMPERATURE= 45.1100946 C

DO YOU WISH TO TRY ANOTHER DESIGN? IF YES,TYPE 1;IF NO,TYPE 2
? 1
WHAT IS THE TUBE DIAMETER IN M?
? .03
WHAT IS THE TUBE LENGTH IN M?
? 20
WATER PROPERTIES AT 50 C FROM DATA BLOCK
WHAT IS THE CALCULATED CAPACITY RATIO?
? .8927
RE= 23402.1043IF RE<2300 SOLUTION IS INVALID
WATER HEAT TRANSFER COEFFICIENT= 2.55464975KW/M↑2 K
```

```
NTU= 2.322391
EFFECTIVENESS 0.725075564

DIAMETER OF TUBE= 3CM
LENGTH OF TUBE= 20M
FLUID OUTLET TEMPERATURE= 42.9909323 C

DO YOU WISH TO TRY ANOTHER DESIGN? IF YES,TYPE 1;IF NO,TYPE 2
? 1
WHAT IS THE TUBE DIAMETER IN M?
? .04
WHAT IS THE TUBE LENGTH IN M?
? 20
WATER PROPERTIES AT 50 C FROM DATA BLOCK
WHAT IS THE CALCULATED CAPACITY RATIO?
? .8927
RE= 17551.5782IF RE<2300 SOLUTION IS INVALID
WATER HEAT TRANSFER COEFFICIENT= 1.52209458KW/M↑2 K

NTU= 2.26621650
EFFECTIVENESS 0.719535099

DIAMETER OF TUBE= 4CM
LENGTH OF TUBE= 20M
FLUID OUTLET TEMPERATURE= 43.6557881 C

DO YOU WISH TO TRY ANOTHER DESIGN? IF YES,TYPE 1;IF NO,TYPE 2
? 2

STOP AT 380
```

Program nomenclature

RO	density
MU	viscosity
K	thermal conductivity of water
PR	Prandtl number
RE	Reynolds number
H	surface heat transfer coefficient
U	overall heat transfer coefficient
L	tube length
A	heat transfer area
NTU	number of transfer units
E	effectiveness
T2	water exit temperature

Program notes

(1) The program allows any combination of diameter and lengths in a single-tube, counter flow heat exchanger to give a desired outlet temperature for a cooled fluid.
(2) The preliminary calculation for the capacity ratio determines the form of the equations (C is always less than 1) and then in lines 150 to 200 the U value is determined, bearing in mind no allowance is made

for change of heat transfer coefficient outside the tube, which could be slightly unrealistic.
(3) In line 220 the value of NTU is found and used to determine the effectiveness (lines 240 to 260) and the outlet temperature in line 280.
(4) This program illustrates the convenience of the NTU method in solving a problem when only inlet temperatures are known together with the flow rate of fluid to be cooled. All the other parameters may be varied but in this problem only tube diameter and length have been used as variable inputs. The simplification of using constant fluid properties at an initially guessed average temperature could be covered by changing this guess or, better, by adjusting the mean temperature in the light of the calculations made.
(5) The problem should be continued to find the minimum temperature that can be achieved by diameter change and then the length can be adjusted to give any required exit temperature. The effect of changing the water flow rate could also be tested.

PROBLEMS

(7.1) Example 7.1 investigated the effect of cavity wall insulation. The same sort of program can be written for flooring, double glazing and loft insulation to determine heat losses, running costs and payback periods. An overall effect for various combinations could also be investigated. Write a suitable program and use the data below to validate the results. Other data may be found in the reference given in example 7.1. An inflation rate for fuel costs could be included if required.

Data
Pitched roof (angle θ) of tiles and roofing felt: $U_r = 1.5$ W/m^2 K
Pitched roof loft air space resistance: $R_a = 0.16$ m^{2}K/W
Uninsulated ceiling resistance: $R_c = 8.2$ m^2 K/W
Thermal conductivity of loft insulation: $k = 0.042$ W/m K (thickness to choice)
Overall upper storey resistance:

$$R = \left(\frac{1}{U_r}\right)\cos\theta + R_a + R_c + R_{\text{insulation}}$$

Intermediate bare wood floors: $U = 1.7$ W/m^2 K
Carpet: $U = 0.06$ W/m^2 K
Underlay: $U = 0.045$ W/m^2 K

Suspended wood floors:	$U = 0.045$ W/m^2 K	use full inside-to-outside
Solid floors:	$U = 0.044$ W/m^2 K	temperature difference
Single glazing:	$U = 5.6$ W/m^2 K	
Double glazing:	$U = 3.0$ W/m^2 K	

(7.2) A pipe 45 m long carries water at 95°C from a heater to a process continuously and is uninsulated in air at an average yearly temperature of 14°C. It is proposed to insulate the pipe and a program is required to relate insulation material and thickness to payback time in years with no cost inflation allowance (the heat losses are made good by auxiliary heating at the process).

Some suitable material data is given below together with other information.

Boiler lagging: $k = 0.07$ W/m K, $\rho = 255$ kg/m^3, £0.25/kg
Insulating powder: $k = 0.06$ W/m K, $\rho = 220$ kg/m^3, £0.30/kg
Rigid glass-fibre: $k = 0.046$ W/mK, $\rho = 160$ kg/m^3. £0.40/kg

The final figure is the installed cost of the insulation. The fuel cost is £0.03/kWh. The pipe outer diameter is 150 mm. The surface heat transfer coefficient for both bare pipe and insulation is 0.7 W/m^2 K. Assume the pipe surface temperature is equal to the water temperature.
(7.3) Example 7.2 determined the temperature distribution in a chimney given the wall temperatures. Allowance may be made for convection at the surfaces by use of the appropriate equations at the surface mesh points. Modify the program of example 7.2 for an inside flue gas temperature of 250°C with an outside air temperature of 15°C. The surface heat transfer coefficient on both inside and outside surfaces is 16 W/m^2 K and the thermal conductivity of brick is 0.9 W/m K. The chimney is 2 m square with a central 0.8 m square flue as in example 7.2.
(7.4) In a heat transfer experiment using a well-insulated smooth tube heated by an electric heating element wound uniformly round the tube, air is blown through the tube (Figure 7.17). The tube has thermocouples situated on the inside wall at a number of stations and the air pressure and temperature are measured at the inlet to the tube. The air

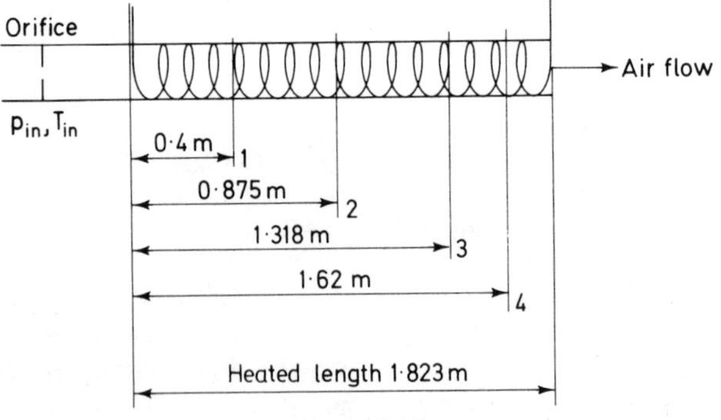

Figure 7.17

flow rate is determined by an orifice plate situated adjacent to these latter measuring points (see problem 3.4).

Write a program to determine and tabulate Re and Nu and then determine the surface heat transfer coefficient at each station for various flow rates using:

(a) $Nu = 0.023\ Re^{0.8}\ Pr^{0.4}$ with the experimental values of Re and Pr;

(b) Reynolds analogy modified by Colburn (obtaining f by Blasius' relation) from the experimental values of Re and Pr; and

(c) The experimental data.

For these latter values a graph should also be drawn to determine a relation of the form

$$Nu = \text{constant}\ Re^{a}$$

to suit this particular heat exchanger.

Experimental data

Net heater input	255 W
Air inlet temperature	28°C
Barometer reading	101 kN/m^2
Orifice diameter	41.3 mm
Orifice coefficient of discharge	0.63
Tube internal diameter	31.76 mm

Air inlet gauge pressure, mm H_2O		470	195	95	43
Orifice pressure drop, mm H_2O		110	36	16	8
Wall temperature	Station 1, °C	37.57	42.63	48.72	57.59
	Station 2, °C	39.13	45.48	52.84	64.09
	Station 3, °C	40.45	47.47	54.23	68.26
	Station 4, °C	41.24	48.60	57.13	70.51

The air temperature at each station is determined by using the steady flow energy equation to equate the heating input between entry and station to the change in enthalpy of the air between entry and the station. The experimental heat transfer coefficient is then given by

$$h_{\text{station}} = \frac{\text{Energy input/unit area}}{(\text{wall temperature} - \text{air temperature})_{\text{station}}}$$

(7.5) In example 7.5 the heat transfer coefficients were obtained from an empirical relation based on dimensional analysis. Repeat the problem, assuming all tubes are smooth, using the Reynolds' analogy modified by Colburn to determine h. Include in the program additional computation:

(a) to tabulate the head loss (h_f) in each design determined from the Blasius' relation for f and the Darcy equation

$$h_f = \frac{4fLV^2}{2gd}$$

(b) to tabulate the mass of each tube package if the density of the tube material is 9000 kg/m^3 and all tube walls are 1.5 mm thick.

(7.6) The heat transfer coefficient to determine the heat flux through the cylinder walls of a reciprocating internal combustion engine may be found from the Eichelberg empirical relation

$$h = 2.1(c)^{1/3} \sqrt{pT} \text{ k cal/m}^2 \text{ hour °C}$$

where c is the mean piston speed in m/s
p is the instantaneous gas pressure in atmospheres
and T is the instantaneous gas temperature in K.

Write a program to determine the heat transfer coefficient in W/m^{2}K and the heat transfer per degree of crank angle (joules) for any crank angle θ. Evaluate the program for an engine with bore 0.0762 m, stroke 0.1111 m and crank length 0.2413 m running at 1750 rev/min with the following data:

Crank angle θ, degrees	*Pressure, kN/m²*	*Temperature, K*
220	85	300
300	313	530
360	4400	1020
390	4700	2800
450	320	900

The area for heat transfer is determined from the piston position, which may be determined from Figure 6.3. Allowance should be made for the area of the cylinder head and the piston by assuming these to be equal to twice the cylinder cross-sectional area.

(7.7) A glasshouse is heated by tubular heaters fitted horizontally. The length of heater is L/m^2 of glass surface area. The tube diameter is D. The energy for heating the glasshouse is supplied by water which has been used as the condenser coolant in an adjacent steam power station. The water gives the tube a surface temperature of 28°C. The minimum outside temperature expected is −10°C and the glasshouse is to be kept at a minimum temperature of 12°C. The U value of the glazing is 6 W/m^2 K.

Write a program to determine L (heater length per m^2 of glass) for various heating tube diameters. Evaluate the program for D = 3, 6, 9 and 12 cm.

For free convection from horizontal tubes:

Laminar flow; $10^4 > Ra > 10^9$, $Nu = 0.525\ (Pr.Gr)^{0.25}$

Turbulent flow; $10^9 > Ra > 10^{12}$, $Nu = 0.129\ (Pr.Gr)^{0.33}$

In these equations film temperature should be used for fluid properties and the characteristic length dimension is D.

(7.8) Write a program to solve two grey body radiation heat transfer problems by the electrical analogy. Test the program for two infinite concentric cylinders each with emissivity 0.2, one of diameter 15 cm and the other of diameter 14 cm. For this problem an analytical solution can be made (see section 7.14) so that the program results can be validated. Choose any convenient values for temperature of the surfaces; for example, imagine the surfaces being those of a vacuum flask holding hot tea or ice.

(7.9) A hall has a heated ceiling at a temperature of 50°C. The ceiling height is 5 m and the hall is 12 m by 16 m. Write a program to determine the heating distribution by radiation at floor level by considering

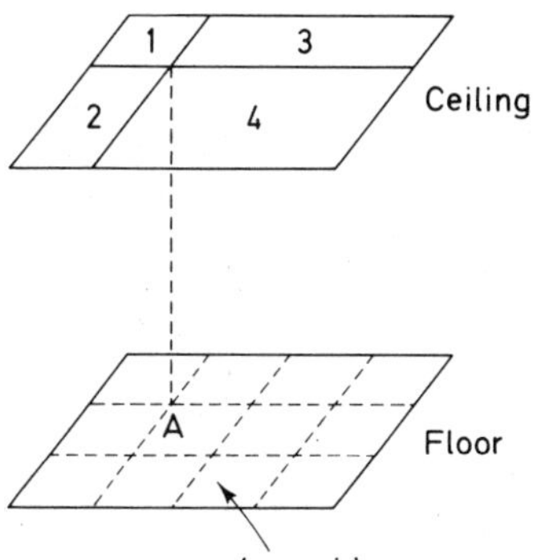

Figure 7.18 Geometric factor diagram

a grid of points 4 m apart (Figure 7.18). The floor temperature is 15°C. Use the Hottel chart (Figure 7.10) adding the geometric factors for each grid point

$$F_A = F_{1A} + F_{2A} + F_{3A} + F_{4A}$$

The emissivity of the ceiling is 0.8 and of the floor is 0.7.

(7.10) Write a program to solve heat transfer by radiation involving three grey bodies. Evaluate the problem for the case of two parallel plates 5 m by 10 m, which are placed 5 m apart in the centre of a very large room (Hottel chart, Figure 7.10).

Data

Plate 1	temperature	500°C
	emissivity	0.2
Plate 2	temperature	250°C
	emissivity	0.5
Room	temperature	20°C

Find the total heat transfer from the hot plate and the total heat transfer to the room.

(7.11) A wedge-shaped fin has a length L and a thickness at the root W tapering to zero thickness at the extreme end. The fin is of unit width. Write a program to determine the heat transfer rate per unit width from the fin using the same data as that in example 7.8. The analytical solution to this problem requires the use of Bessel functions to solve the differential equation and need not be attempted.

The following 'examples' are not precisely presented but represent ideas for other problems which would be suitable for simple investigations.

(7.12) Analysis of laboratory tests.
(7.13) Pitch of fins: spacing versus length for optimising heat transfer.
(7.14) Car heater performance: output temperature for various flow rates; total heat flux.
(7.15) Heating costs by electricity, oil, gas, coal, etc.
(7.16) Performance of rotary regenerators (heat wheels).
(7.17) Solar panel performance in terms of sunshine hours per month versus 'degree days' representing heating needs per month for various areas of the United Kingdom.
(7.18) Try introducing some inflation rates and capital servicing costs into payback period calculations. Incorporate any accounting techniques available.
(7.19) Make a study of your own home heating problems and try to find an optimum answer. If you can apply the results, do so and see how close the predictions are.
(7.20) Condensation in the fabric of walls.
(7.21) Annular fins.
(7.22) Three-dimensional steady state conduction problems.
(7.23) One-dimensional transient conduction problems.
(7.24) Temperature distribution and conduction in conductors carrying electric current.

Index